T0344593

# Nanoscale Thermoelectric Materials: Thermal and Electrical Transport, and Applications to Solid-State Cooling and Power Generation

MATERIALS RESEARCH SOCIETY
SYMPOSIUM PROCEEDINGS VOLUME 1543

# Nanoscale Thermoelectric Materials: Thermal and Electrical Transport, and Applications to Solid-State Cooling and Power Generation

Symposia held April 1–5, 2013, San Francisco, California U.S.A.

EDITORS

**Scott P. Beckman**
Iowa State University
Ames, Iowa, U.S.A.

**Harald Böttner**
Fraunhofer-Institut für Physikalische
Messtechnik IPM
Freiburg. Germany

**Yann Chalopin**
Ecole Centrale Paris
Chatenay Malabry, France

**Christopher Dames**
University of California, Berkeley
Berkeley, California, U.S.A.

**P. Alex Greaney**
Oregon State University
Corvallis, Oregon, U.S.A.

**Patrick Hopkins**
University of Virginia
Charlottesville, Virginia, U.S.A.

**Baowen Li**
National University of Singapore
Singapore, Singapore

**Tako Mori**
National Institute for Materials Science
Tsukuba, Japan

**Takeshi Nishimatsu**
Tohoku University
Sendai, Japan

**Kevin Pipe**
University of Michigan
Ann Arbor, Michigan, U.S.A.

**Rama Venkatasubramanian**
Johns Hopkins University
Laurel, Maryland, U.S.A.

Materials Research Society
Warrendale, Pennsylvania

Shaftesbury Road, Cambridge CB2 8EA, United Kingdom

One Liberty Plaza, 20th Floor, New York, NY 10006, USA

477 Williamstown Road, Port Melbourne, VIC 3207, Australia

314–321, 3rd Floor, Plot 3, Splendor Forum, Jasola District Centre, New Delhi – 110025, India

103 Penang Road, #05–06/07, Visioncrest Commercial, Singapore 238467

Cambridge University Press is part of Cambridge University Press & Assessment, a department of the University of Cambridge.

We share the University's mission to contribute to society through the pursuit of education, learning and research at the highest international levels of excellence.

www.cambridge.org
Information on this title: www.cambridge.org/9781605115207

Materials Research Society
506 Keystone Drive, Warrendale, PA 15086
http://www.mrs.org



First published 2013

CODEN: MRSPDH

*A catalogue record for this publication is available from the British Library*

ISBN 978-1-605-11520-7 Hardback

# CONTENTS

**Preface** . . . . . . . . . . . . . . . . . . . . . . . . . . . . . . . . . . . . . . . . . . . . . . . . . **ix**

**Materials Research Society Symposium Proceedings** . . . . . . . . . . . . . . . **xi**

*NEW MATERIALS APPROACHES*

**A New Thermoelectric Concept Using Large Area PN Junctions** . . . . . . . **3**
R. Chavez, A. Becker, V. Kessler, M. Engenhorst, N. Petermann, H. Wiggers, G. Schierning, and R. Schmechel

* **Thermoelectric Transport in Topological Insulator $Bi_2Te_2Se$ Bulk Crystals** . . . . . . . . . . . . . . . . . . . . . . . . . . . . . . . . . . . . . . . . . . . . . . . . . **9**
Yang Xu, Helin Cao, Ireneusz Miotkowski, and Yong P. Chen

**Large Electrocaloric Effect from Electrical Field Induced Orientational Order-disorder Transition in Nematic Liquid Crystals Possessing Large Dielectric Anisotropy** . . . . . . . . . . . . . . . . . . . **13**
Xiao-Shi Qian, S.G. Lu, Xinyu Li, Haiming Gu, L.-C. Chien, and Q.M. Zhang

*MODELING/THEORY*

* **First-principles Investigations on the Thermoelectric Properties of $Bi_2Te_3$ Doped with Se** . . . . . . . . . . . . . . . . . . . . . . . . . . . . . . . . . . . **23**
Liwen F. Wan and Scott P. Beckman

* **Giant Thermoelectric Effect in Graded Micro-nanoporous Materials** . . . . . . . . . . . . . . . . . . . . . . . . . . . . . . . . . . . . . . . . . . . . . . . . . **29**
Dimitrios G. Niarchos, Roland H. Tarkhanyan, and Alexandra Ioannidou

*Invited Paper

**Computational Modeling the Electrocaloric Effect for Solid-state Refrigeration** . . . . . . . . . . . . . . . . . . . . . . . . . . . . . . . . . . . . . . . . . . . . . .39
J.A. Barr, T. Nishimatsu, and S.P. Beckman

**Heat Transport between Heat Reservoirs Mediated by Quantum Systems** . . . . . . . . . . . . . . . . . . . . . . . . . . . . . . . . . . . . . . . . . . . . .43
George Y. Panasyuk, George A. Levin, and Kirk L. Yerkes

**Detailed Theoretical Investigation and Comparison of the Thermal Conductivities of n- and p-type $Bi_2Te_3$ Based Alloys** . . . . . . . . . . . . . . .49
Ö. Ceyda Yelgel and Gyaneshwar P. Srivastava

## *LAYERED STRUCTURES*

**Enhanced Thermoelectric Properties of Al-doped ZnO Thin Films.** . . . .57
P. Mele, S. Saini, H. Abe, H. Honda, K. Matsumoto, K. Miyazaki, H. Hagino, and A. Ichinose

**Mapping Thermal Resistance Around Vacancy Defects in Graphite** . . . . . . . . . . . . . . . . . . . . . . . . . . . . . . . . . . . . . . . . . . . . . .65
Laura de Sousa Oliveira and P. Alex Greaney

**Thermal Conductivity of Regularly Spaced Amorphous/Crystalline Silicon Superlattices. A Molecular Dynamics Study** . . . . . . . . . . . . . . . .71
Konstantinos Termentzidis, Arthur France-Lanord, Etienne Blandre, Tristan Albaret, Samy Merabia, Valentin Jean, and David Lacroix

## *NANOSTRUCTURED BULK AND COMPOSITES*

* **Thermoelectric Properties of Ru and In Substituted Misfit-layered $Ca_3Co_4O_9$** . . . . . . . . . . . . . . . . . . . . . . . . . . . . . . . . . . . . . .83
Gesine Saucke, Sascha Populoh, Nina Vogel-Schäuble, Leyre Sagarna, Kailash Mogare, Lassi Karvonen, and Anke Weidenkaff

*Invited Paper

**Enhanced Thermoelectric Figure-of-merit at Room Temperature in Bulk Bi(Sb)Te(Se) With Grain Size of ~100nm** . . . . . . . . . . . . . . . . . . . .93
Tsung-ta E. Chan, Rama Venkatasubramanian, James M. LeBeau, Peter Thomas, Judy Stuart, and Carl C. Koch

**Impact of Rapid Thermal Annealing on Thermoelectric Properties of Bulk Nanostructured Zinc Oxide** . . . . . . . . . . . . . . . . . . . . . . . . . . . .99
Markus Engenhorst, Devendraprakash Gautam, Carolin Schilling, Markus Winterer, Gabi Schierning, and Roland Schmechel

**Fabrication and Characterization of Nanostructured Bulk Skutterudites** . . . . . . . . . . . . . . . . . . . . . . . . . . . . . . . . . . . . . . . . .105
Mohsen Y. Tafti, Mohsin Saleemi, Alexandre Jacquot, Martin Jägle, Mamoun Muhammed, and Muhammet S. Toprak

## *NANOWIRES, NANOTUBES, AND NANOCRYSTALS*

**A Novel Approach to Synthesize Lanthanum Telluride Thermoelectric Thin Films in Ambient Conditions** . . . . . . . . . . . . . . . . . . . . . . . . . . . .113
Su (Ike) Chih Chi, Stephen L. Farias, and Robert C. Cammarata

**Fabrication of Thermally-conductive Carbon Nanotubes-copper Oxide Heterostructures** . . . . . . . . . . . . . . . . . . . . . . . . . . . . . . . . . . . .119
Yuan Li and Nitin Chopra

**Enhanced Thermopower of GaN Nanowires with Transitional Metal Impurities** . . . . . . . . . . . . . . . . . . . . . . . . . . . . . . . . . . . . . . . . . .125
G.A. Nemnes, Camelia Visan, T.L. Mitran, Adela Nicolaev, L. Ion, and S. Antohe

**Growth of Polycrystalline Indium Phosphide Nanowires on Copper** . . . . . . . . . . . . . . . . . . . . . . . . . . . . . . . . . . . . . . . . . . . . .131
Kate J. Norris, Junce Zhang, David M. Fryauf, Elane Coleman, Gary S. Tompa, and Nobuhiko P. Kobayashi

**Quaternary Chalcogenide Nanocrystals: Synthesis of $Cu_2ZnSnSe_4$ by Solid State Reaction and their Thermoelectric Properties** . . . . . . . . . . .137
Umme Farva and Chan Park

**Design and Evaluation of Carbon Nanotube Based Nanofluids for Heat Transfer Applications** ............................143
Sathya P. Singh, Nader Nikkam, Morteza Ghanbarpour, Muhammet S. Toprak, M. Muhammed, and Rahmatollah Khodabandeh

## *NANOSCALE TRANSPORT PHENOMENA*

**Solute Effects on Interfacial Thermal Conductance**...............151
Andrew J. Green and Hugh H. Richardson

**Heat Transfer between a Hot AFM Tip and a Cold Sample: Impact of the Air Pressure**..........................................159
Pierre-Olivier Chapuis, Emmanuel Rousseau, Ali Assy, Séverine Gomès, Stéphane Lefèvre, and Sebastian Volz

**Thermal and Rheological Properties of Micro- and Nanofluids of Copper in Diethylene Glycol – as Heat Exchange Liquid** ...........165
Nader Nikkam, Morteza Ghanbarpour, Mohsin Saleemi, Muhammet S. Toprak, Mamoun Muhammed, and Rahmatollah Khodabandeh

**Carrier Mapping in Thermoelectric Materials**....................171
Georgios S. Polymeris, Euripides Hatzikraniotis, Eleni C. Stefanaki, Eleni Pavlidou, Theodora Kyratsi, Konstantinos M. Paraskevopoulos, and Mercouri G. Kanatzidis

**Author Index** ..............................................177

**Subject Index** ..............................................179

# PREFACE

At the Material Research Society Spring 2013 meeting, held in San Francisco April 1-5, 2013, three symposia were held that focused on thermal-to-electric energy conversion and thermal transport: Symposium H: Nanoscale Thermoelectrics—Materials and Transport Phenomena – II, Symposium I: Materials for Solid-State Refrigeration, and Symposium V: Nanoscale Heat Transport—From Fundamentals to Devices. Although each technical session was focused on a different aspect of this subject, the intellectual commensurability of these symposia warranted the publication of their proceedings in a single volume.

The sessions in symposium H covered the latest approaches and results in nanoscale thermoelectric materials and their devices for thermal-to-electric conversion, energy harvesting as well as for high-performance cooling and thermal management. Recent advances, including nanoscale materials to reduce lattice thermal conductivity without significantly affecting electronic transport, resonant states, delta-doping, topological insulators, energy filtering to control the flow of desirable heat-transporting carriers and organic/molecular structures were discussed. The sessions, which included several invited talks from leading experts, highlighted the fundamentals of nanoscale materials synthesis and first-principles calculations aimed at understanding the nanoscale physics of transport. The proceedings of the symposium is a collection of papers that highlight the multidisciplinary nature of the materials research, including measurement methods and characterization tools for thermal and electrical interfaces.

The sessions in symposium I covered the latest approaches in materials used for solid-state refrigeration technologies. The development of this symposium was in part driven by interests in technologies to replace the existing vapor-compression cycle devices, which are notoriously inefficient. It was also motivated by recent interest in technologies aimed at micro-scale refrigeration devices, e.g., cooling on a chip. In addition to presentations on thermoelectric solid-state technologies, which have experienced recent advances due to the development of nanostructured materials, presentations about new approaches were given including devices based on the magnetocaloric effect, the electrocaloric effect, and the elastocaloric effect. There are also common challenges to solid-state refrigeration that are shared by all phenomenological approaches; for example, the control and optimization of thermal conductivity at and around the operating region. These issues were also discussed in this session. By creating sessions that included researchers from diverse backgrounds we hoped to stimulate discussions across fields and provide a receptive forum for new concepts in refrigeration.

The sessions in symposium V covered the latest approaches in heat transport within nanostructured materials, nanoscale devices, and lithographically-defined nanostructures, where structural length scales overlap with intrinsic phonon and electron length scales, leading to the strong modification of thermal transport mechanisms. Topics of invited and contributed presentations included phonon barriers and superlattices, nanowires, graphene, carbon nanotubes and carbon nanotube composites, electron-phonon interactions, organic and hybrid materials, thermal radiation, microfluidics, and phase change materials. Emerging measurement and simulation techniques were highlighted, as well as applications to medical therapies, thermoelectric energy conversion, and energy storage. Profs. Ali Shakouri and Alan McGaughey, from Purdue University and Carnegie Mellon University respectively, gave well-attended tutorials on experimental and computational methods for studying nanoscale heat transport.

Scott P. Beckman
Harald Böttner
Yann Chalopin
Christopher Dames
P. Alex Greaney
Patrick Hopkins
Baowen Li
Tako Mori
Takeshi Nishimatsu
Kevin Pipe
Rama Venkatasubramanian

September 2013

# MATERIALS RESEARCH SOCIETY SYMPOSIUM PROCEEDINGS

Volume 1536 — Film Silicon Science and Technology, 2013, P. Stradins, A. Boukai, F. Finger, T. Matsui, N. Wyrsch, ISBN 978-1-60511-513-9
Volume 1537E — Organic and Hybrid Photovoltaic Materials and Devices, 2013, S.W. Tsang, ISBN 978-1-60511-514-6
Volume 1538 — Compound Semiconductors: Thin-Film Photovoltaics, LEDs, and Smart Energy Controls, 2013, M. Al-Jassim, C. Heske, T. Li, M. Mastro, C. Nan, S. Niki, W. Shafarman, S. Siebentritt, Q. Wang, ISBN 978-1-60511-515-3
Volume 1539E — From Molecules to Materials—Pathways to Artificial Photosynthesis, 2013, J.-H. Guo, ISBN 978-1-60511-516-0
Volume 1540E — Materials and Integration Challenges for Energy Generation and Storage in Mobile Electronic Devices, 2013, M. Chhowalla, S. Mhaisalkar, A. Nathan, G. Amaratunga, ISBN 978-1-60511-517-7
Volume 1541E — Materials for Vehicular and Grid Energy Storage, 2013, J. Kim, ISBN 978-1-60511-518-4
Volume 1542E — Electrochemical Interfaces for Energy Storage and Conversion—Fundamental Insights from Experiments to Computations, 2013, J. Cabana, ISBN 978-1-60511-519-1
Volume 1543 — Nanoscale Thermoelectric Materials, Thermal and Electrical Transport, and Applications to Solid-State Cooling and Power Generation, 2013, S.P. Beckman, H. Böttner, Y. Chalopin, C. Dames, P.A. Greaney, P. Hopkins, B. Li, T. Mori, T. Nishimatsu, K. Pipe, R. Venkatasubramanian, ISBN 978-1-60511-520-7
Volume 1544E — *In-Situ* Characterization Methods in Energy Materials Research, 2014, J.D. Baniecki, P.C. McIntyre, G. Eres, A.A. Talin, A. Klein, ISBN 978-1-60511-521-4
Volume 1545E — Materials for Sustainable Development, 2013, R. Pellenq, ISBN 978-1-60511-522-1
Volume 1546E — Nanoparticle Manufacturing, Functionalization, Assembly and Integration, 2013, H. Fan, T. Hyeon, Z. Tang, Y. Yin, ISBN 978-1-60511-523-8
Volume 1547 — Solution Synthesis of Inorganic Functional Materials—Films, Nanoparticles and Nanocomposites, 2013, M. Jain, Q.X. Jia, T. Puig, H. Kozuka, ISBN 978-1-60511-524-5
Volume 1548E — Nanomaterials in the Subnanometer-Size Range, 2013, J.S. Martinez, ISBN 978-1-60511-525-2
Volume 1549 — Carbon Functional Nanomaterials, Graphene and Related 2D-Layered Systems, 2013, P.M. Ajayan, J.A. Garrido, K. Haenen, S. Kar, A. Kaul, C.J. Lee, J.A. Robinson, J.T. Robinson, I.D. Sharp, S. Talapatra, R. Tenne, M. Terrones, A.L. Elias, M. Paranjape, N. Kharche, ISBN 978-1-60511-526-9
Volume 1550E — Surfaces of Nanoscale Semiconductors, 2013, M.A. Filler, W.A. Tisdale, E.A. Weiss, R. Rurali, ISBN 978-1-60511-527-6
Volume 1551 — Nanostructured Semiconductors and Nanotechnology, 2013, I. Berbezier, J-N. Aqua, J. Floro, A. Kuznetsov,ISBN 978-1-60511-528-3
Volume 1552 — Nanostructured Metal Oxides for Advanced Applications, 2013, A. Vomiero, F. Rosei, X.W. Sun, J.R. Morante, ISBN 978-1-60511-529-0
Volume 1553E — Electrical Contacts to Nanomaterials and Nanodevices, 2013, F. Léonard, C. Lavoie, Y. Huang, K. Kavanagh, ISBN 978-1-60511-530-6
Volume 1554E — Measurements of Atomic Arrangements and Local Vibrations in Nanostructured Materials, 2013, A. Borisevich, ISBN 978-1-60511-531-3
Volume 1556E — Piezoelectric Nanogenerators and Piezotronics, 2013, X. Wang, C. Falconi, S-W. Kim, H.A. Sodano, ISBN 978-1-60511-533-7
Volume 1557E — Advances in Scanning Probe Microscopy for Imaging Functionality on the Nanoscale, 2013, S. Jesse, H.K. Wickramasinghe, F.J. Giessibl, R. Garcia, ISBN 978-1-60511-534-4
Volume 1558E — Nanotechnology and Sustainability, 2013, L. Vayssieres, S. Mathur, N.T.K. Thanh, Y. Tachibana, ISBN 978-1-60511-535-1
Volume 1559E — Advanced Interconnects for Micro- and Nanoelectronics—Materials, Processes and Reliability, 2013, E. Kondoh, M.R. Baklanov, J.D. Bielefeld, V. Jousseaume, S. Ogawa, ISBN 978-1-60511-536-8
Volume 1560E — Evolutions in Planarization—Equipment, Materials, Techniques and Applications, 2013, C. Borst, D. Canaperi, T. Doi, J. Sorooshian, ISBN 978-1-60511-537-5
Volume 1561E — Gate Stack Technology for End-of-Roadmap Devices in Logic, Power and Memory, 2013, S. Banerjee,ISBN 978-1-60511-538-2

# MATERIALS RESEARCH SOCIETY SYMPOSIUM PROCEEDINGS

Volume 1562E — Emerging Materials and Devices for Future Nonvolatile Memories, 2013, Y. Fujisaki, P. Dimitrakis, D. Chu, D. Worledge, ISBN 978-1-60511-539-9
Volume 1563E — Phase-Change Materials for Memory, Reconfigurable Electronics, and Cognitive Applications, 2013, R. Calarco, P. Fons, B.J. Kooi, M. Salinga, ISBN 978-1-60511-540-5
Volume 1564E — Single-Dopant Semiconductor Optoelectronics, 2014, M.E. Flatté, D.D. Awschalom, P.M. Koenraad, ISBN 978-1-60511-541-2
Volume 1565E — Materials for High-Performance Photonics II, 2013, T.M. Cooper, S.R. Flom, M. Bockstaller, C. Lopes, ISBN 978-1-60511-542-9
Volume 1566E — Resonant Optics in Metallic and Dielectric Structures—Fundamentals and Applications, 2013, L. Cao, N. Engheta, J. Munday, S. Zhang, ISBN 978-1-60511-543-6
Volume 1567E — Fundamental Processes in Organic Electronics, 2013, A.J. Moule, ISBN 978-1-60511-544-3
Volume 1568E — Charge and Spin Transport in Organic Semiconductor Materials, 2013, H. Sirringhaus, J. Takeya, A. Facchetti, M. Wohlgenannt, ISBN 978-1-60511-545-0
Volume 1569 — Advanced Materials for Biological and Biomedical Applications, 2013, M. Oyen, A. Lendlein, W.T. Pennington, L. Stanciu, S. Svenson, ISBN 978-1-60511-546-7
Volume 1570E — Adaptive Soft Matter through Molecular Networks, 2013, R. Ulijn, N. Gianneschi, R. Naik, J. van Esch, ISBN 978-1-60511-547-4
Volume 1571E — Lanthanide Nanomaterials for Imaging, Sensing and Optoelectronics, 2013, H. He, Z-N. Chen, N. Robertson, ISBN 978-1-60511-548-1
Volume 1572E — Bioelectronics—Materials, Interfaces and Applications, 2013, A. Noy, N. Ashkenasy, C.F. Blanford, A. Takshi, ISBN 978-1-60511-549-8
Volume 1574E — Plasma and Low-Energy Ion-Beam-Assisted Processing and Synthesis of Energy-Related Materials, 2013, G. Abrasonis, ISBN 978-1-60511-551-1
Volume 1575E — Materials Applications of Ionic Liquids, 2013, D. Jiang, ISBN 978-1-60511-552-8
Volume 1576E — Nuclear Radiation Detection Materials, 2014, A. Burger, M. Fiederle, L. Franks, D.L. Perry, ISBN 978-1-60511-553-5
Volume 1577E — Oxide Thin Films and Heterostructures for Advanced Information and Energy Technologies, 2013, G. Herranz, H-N. Lee, J. Kreisel, H. Ohta, ISBN 978-1-60511-554-2
Volume 1578E — Titanium Dioxide—Fundamentals and Applications, 2013, A. Selloni , ISBN 978-1-60511-555-9
Volume 1579E — Superconducting Materials—From Basic Science to Deployment, 2013, Q. Li, K. Sato, L. Cooley, B. Holzapfel, ISBN 978-1-60511-556-6
Volume 1580E — Size-Dependent and Coupled Properties of Materials, 2013, B.G. Clark, D. Kiener, G.M. Pharr, A.S. Schneider, ISBN 978-1-60511-557-3
Volume 1581E — Novel Functionality by Reversible Phase Transformation, 2013, R.D. James, S. Fähler, A. Planes, I. Takeuchi, ISBN 978-1-60511-558-0
Volume 1582E — Extreme Environments—A Route to Novel Materials, 2013, A. Goncharov, ISBN 978-1-60511-559-7
Volume 1583E — Materials Education—Toward a Lab-to-Classroom Initiative, 2013, E.M. Campo, C.C. Broadbridge, K. Hollar, C. Constantin, ISBN 978-1-60511-560-3

**Prior Materials Research Symposium Proceedings available by contacting Materials Research Society**

# New Materials Approaches

**Mater. Res. Soc. Symp. Proc. Vol. 1543 © 2013 Materials Research Society**
**DOI: 10.1557/opl.2013.954**

# A new thermoelectric concept using large area PN junctions

R. Chavez, A. Becker, V. Kessler, M. Engenhorst, N. Petermann, H. Wiggers, G. Schierning, R. Schmechel
Faculty of Engineering and Center for NanoIntegration Duisburg-Essen (CENIDE), University of Duisburg-Essen, 47057 Duisburg, Germany

## ABSTRACT

A new thermoelectric concept using large area silicon PN junctions is experimentally demonstrated. In contrast to conventional thermoelectric generators where the n-type and p-type semiconductors are connected electrically in series and thermally in parallel, we demonstrate a large area PN junction made from densified silicon nanoparticles that combines thermally induced charge generation and separation in a space charge region with the conventional Seebeck effect by applying a temperature gradient parallel to the PN junction. In the proposed concept, the electrical contacts are made at the cold side eliminating the need for contacts at the hot side allowing temperature gradients greater than 100K to be applied. The investigated PN junction devices are produced by stacking n-type and p-type nanopowder prior to a densification process. The nanoparticulate nature of the densified PN junction lowers thermal conductivity and increases the intraband traps density which we propose is beneficial for transport across the PN junction thus enhancing the thermoelectric properties. A fundamental working principle of the proposed concept is suggested, along with characterization of power output and output voltages per temperature difference that are close to those one would expect from a conventional thermoelectric generator.

## INTRODUCTION

In recent years non-fossil fuels and efficiency of energy conversion have gained great interest as the world prepares to fulfill the projected energy demand. In regards to the later, thermoelectricity could play an important role by allowing a practical form of heat waste recovery as nearly 60% of the energy consumed in the U.S.A. is wasted in the form of heat[1]. Although efforts are being done to incorporate thermoelectric generators (TEGs) in automobiles, among other difficulties, the relatively low efficiencies, usually between 5-15% [1], have refrained the application of thermoelectric generators to deep space missions [2,3]. However, the reduction of the thermal conductivity in nanostructured materials has brought improvements to the thermoelectric material's figure of merit [4–6] and thus the efficiency of the energy conversion process defined by the operation temperatures and the figure of merit which is usually written as:

$$zT = \frac{\sigma\alpha^2 T}{\kappa} \quad (1)$$

where T is the temperature, σ is the electrical conductivity, α the Seebeck coefficient and κ the thermal conductivity.

Conventional thermoelectric generators consist of p and n-type semiconductor materials thermally connected in parallel and electrically connected in series, where a heat source is applied on one side and part of the thermal flux is converted into electrical energy by means of the Seebeck effect (Figure 1a).

Span et. al. have proposed a new approach to the construction and operation of thermoelectric thin film generators in which a large area PN junction is used not only to mechanically hold the p and n-type semiconductors but to enhance the thermoelectric process by thermal generation and separation of electron hole pairs (EHPs) [7–10]. While originally proposed by Span et. al. as a thin film device, it is here transferred to bulk structures. This also removes the difficulties involved with high temperature electrical contacts and the ceramic substrate at the hot side removing the temperature drop across the substrate (Figure 1b).

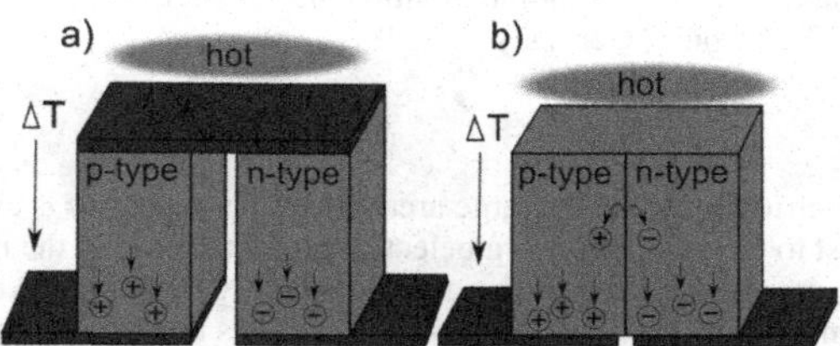


**Figure 1.** Comparison between a conventional thermoelectric generator and a new concept, originally proposed for thin films TEGs, using large area PN junctions where the contacts at the hot side are not needed.

Although investigation on thermoelectric PN junctions has been reported in [11,12], it was not done in the context proposed by Span et. al. Hence the aim of this paper is to demonstrate that a thermoelectric generator could be built using nanostructured silicon PN junctions which offer the advantage of high temperature operation.

**EXPERIMENT**

Doped silicon nanoparticles were produced in a bottom up approach starting with a gaseous precursor in a microwave plasma reactor as described in [13]. The doping concentration as calculated from the gas mixture in the plasma reactor is in the order of $10^{20}$ $cm^{-3}$. To prepare the PN junctions, a layer of p-type (Boron doped) and a layer of n-type (Phosphorus doped) nanoparticles were densified together in a current assisted sintering process in which Joule heating partially melts the nanoparticles while pressure is applied to the stacked nanoparticles layers thus compacting the nanoparticles into a solid PN junction. Because Peltier effects which are proportional to the electrical current should not be ignored during the compaction process in which kilo-Amperes are forced through the semiconductor material, one PN junction is prepared in 'forward bias', and a second sample in 'reverse bias' by inverting the stacking order of the nanoparticles powder. Details about the preparation of such PN junctions can be found in [14,15].

The cut densified PN junctions are shown in Figure 2b. The thermoelectric properties are measured in an apparatus schematically depicted in Figure 2a where the temperature gradient is applied parallel to the PN interface with a heating resistor. The cold side is actively cooled with water flowing through the steel sample holder, and mechanical and electrical contacts are made there at the cold side. The temperatures are measured with thermocouples. From this experimental setup, power output, open circuit voltage and short circuit current per temperature difference are characterized.

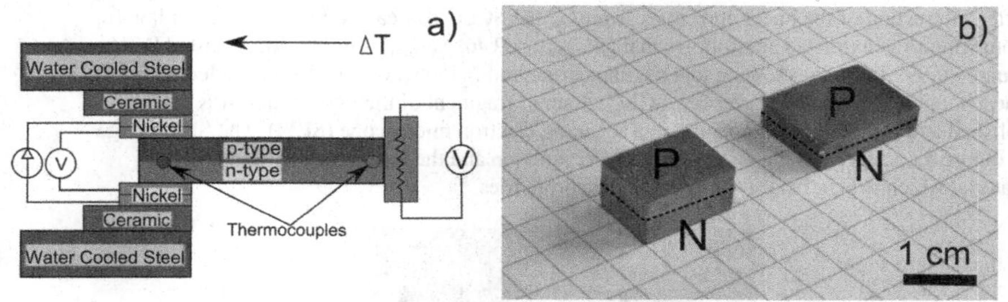


**Figure 2.** a) Schematic (not to scale) of the measuring setup. b) Measured samples: left, PN Forward, right, PN Reverse. The dotted lines indicate the junction.

## RESULTS

First, the electronic properties are summarized. The 4-wire IV characteristics in Figure 3a were obtained at room temperature under thermal equilibrium and dark conditions. Nearly no blocking behavior is observed as opposed to the expectation from a PN junction; this is due to a combination of a number of reasons. For example, during preparation of the PN junction, the n and p-type powders are manually pre-compacted and this creates a great degree of powder intermixing and charge carrier compensation. Moreover, during sintering the partially molten nanoparticles mix further and this can explain why the carrier concentration is lower in the compacted nanoparticles compared to the nominal carrier concentration in the powder [14]. For these reasons, mid bandgap states are formed and this explains the non-blocking behavior of the PN junction in consistency with a Space Charge Limited Current (SCLC) model in which intrinsic shunt paths develop a power dependency of the current from voltage [16]:

$$|I_{SCL}| \propto |V|^{\gamma+1} \quad (2)$$

where $\gamma > 0$ depends on the exact nature of the trap distribution inside the bandgap.

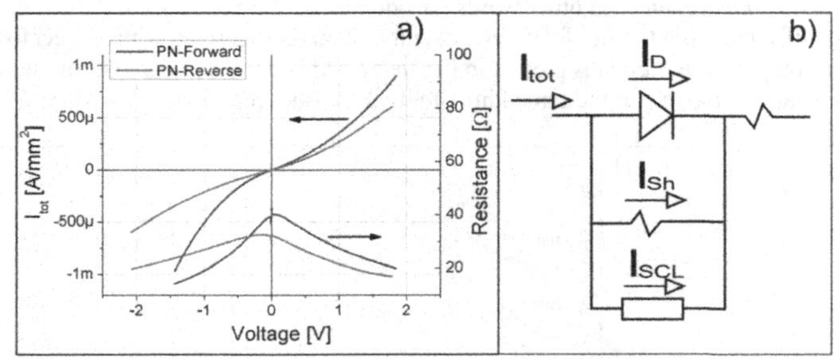


**Figure 3 – a) IV Characteristic of the nanostructured PN junctions. The average resistance shows the poor blocking behavior. b) Equivalent circuit of the PN junctions.**

Analysis of the equivalent circuit in Figure 3b shows that virtually all the current bypasses the diode, that is $I_{tot} \approx I_{Sh} + I_{SCL}$ as no blocking or exponential behavior is observed.

After subtraction of the linear current $I_{Sh}$ from $I_{tot}$, the SCLC $I_{SCL}$ can be fitted to a power law fit, from which $\gamma$ is extracted and follows that $\gamma_{PN\text{-}Forward}=0.46$, $\gamma_{PN\text{-}Reverse}=0.39$, with adjusted R square of .999 for both, showing that the nanostructured PN junctions are best modeled with a linear shunt resistor and a power law SCLC. A small fragment of the PN junction was cut, polished and etched prior to imaging in a scanning electron microscope (SEM). The SEM image in Figure 4 corroborates the poorly defined PN junction and the nanostructure nature of the material which is beneficial for thermoelectric properties.

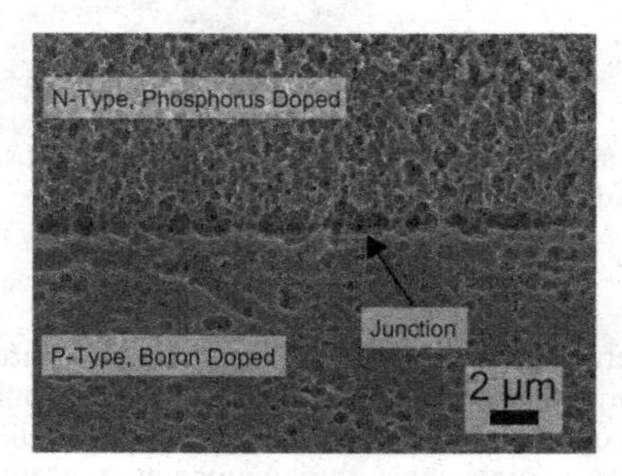


**Figure 4 –SEM image shows the poorly defined PN junction and the nanostructured nature of the material.**

The thermoelectric properties are investigated using the apparatus in depicted in Figure 2a. The absence of electrical contacts at the hot side allows for temperature gradients larger than 140°C to be applied (the limitation being heating power). Figure 5a shows the IV characteristic from the 'PN Reverse' sample under different temperature gradients. From these IV characteristics the open circuit voltage and short circuit current can be extracted as well as the power generation and internal resistance. Let us first consider the open circuit voltage per temperature difference which closely matches in the samples regardless of preparation (Figure 5b). A coefficient of *Volts/Kelvin* or in other words the equivalent of a Seebeck coefficient is approximated by differentiating the fit of the open circuit voltage in Figure 5b with respect to $\Delta T$. The coefficient of output *Volts/Kelvin* is plotted in Figure 5c and is close compared to the sum of p and n type unipolar coefficients of the same kind of powder earlier reported [1718].

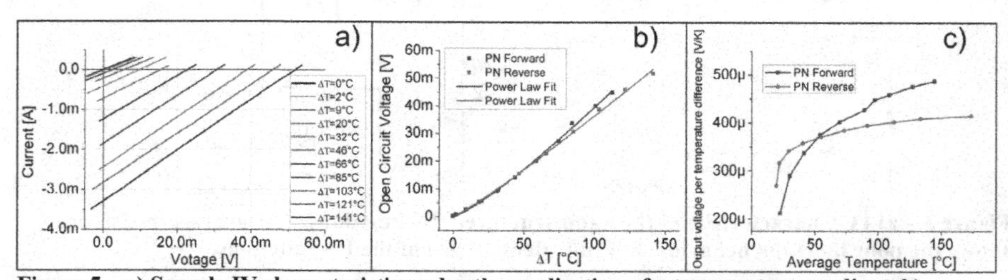


**Figure 5 – a) Sample IV characteristic under the application of a temperature gradient, b) device open circuit voltage, c) differentiation of the fits in Figure 5b yields an equivalent output voltage per temperature difference (Seebeck coefficient equivalent).**

The total power output and internal resistance can be extracted from Figure 5a, and the maximum power output is plotted in Figure 6a. Because the samples differ in dimension, the power output is normalized by the area and plotted in Figure 6b where it can be seen that regardless of sample preparation, the power generation per unit area matches closely between the two samples. The relatively high internal resistance (Figure 5c) is a considerable limitation in the electrical power output, however, interestingly and opposed to conventional generators, the internal resistance decreases with temperature. This as explained by conventional diode theory, would be due to thermal generation of intrinsic carriers and electron-hole pair separation within the small space charge region domains that might exist because an SCLC model is not expected to be strongly dependent on temperature[16].

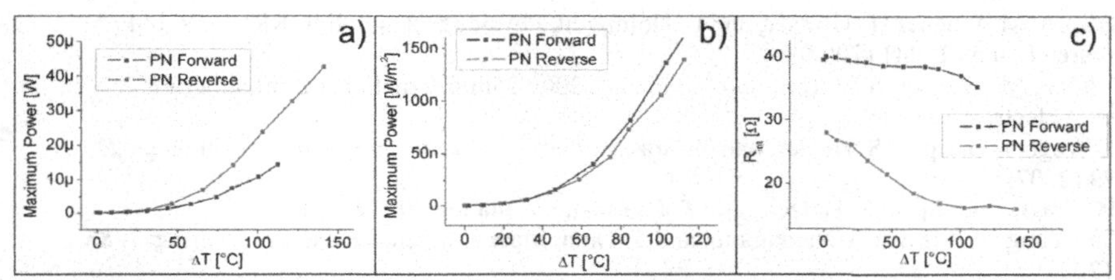


**Figure 6 – a) Total power output of a single pair PN thermoelectric generator (PN-TEG). b) Power output normalized by the PN area. c) A decrease of the internal resistance is observed opposite to conventional generators.**

Commercial thermoelectric generators or Peltier coolers normally have about 100 pairs for which the concept we propose would produce 4.5mW at $\Delta T \approx 145°C$, this in consideration that $\Delta T$ at the present is limited by heating power and not by material stability related issues. The maximum temperature at the hot side is ≈220°C. It is worth noting that in simulations, the PN TEGs are predicted to outperform conventional TEGs for $\Delta T > 300°C$ and that point has not been yet reached [10].

## CONCLUSIONS

We demonstrate an alternative concept towards the construction and operation of thermoelectric generators in which a nanostructured PN junction is used to make mechanical and electrical contacts. This makes it possible to apply large temperature gradients > 140°C which are limited by heating power. Power output is characterized along with electronic properties and demonstrates that such an approach is feasible for the construction of high temperature TEGs. Evaluation of the performance of the PN-TEGs and comparison with conventional technologies is due as the main goal here is to demonstrate functionality.

## ACKNOWLEDGMENTS

Financial support in the frame of a young investigator grant by the Ministry for innovation, science and research of the State North Rhine Westphalia in Germany is gratefully acknowledged.

## REFERENCES

[1] A. Shakouri, Annual Review of Materials Research **41**, 399 (2011).
[2] G.L. Bennett, 26 (2006).
[3] G.L. Bennett and J.J. Lombardo, 26 (2006).
[4] S.K. Bux, R.G. Blair, P.K. Gogna, H. Lee, G. Chen, M.S. Dresselhaus, R.B. Kaner, and J.-P. Fleurial, Advanced Functional Materials **19**, 2445 (2009).
[5] P. Pichanusakorn and P. Bandaru, Materials Science and Engineering: R: Reports **67**, 19 (2010).
[6] C.J. Vineis, A. Shakouri, A. Majumdar, and M.G. Kanatzidis, Advanced Materials (Deerfield Beach, Fla.) **22**, 3970 (2010).
[7] G. Span, M. Wagner, T. Grasser, and L. Holmgren, Physica Status Solidi (RRL) – Rapid Research Letters **1**, 241 (2007).
[8] G. Span, M. Wagner, S. Holzer, and T. Grasser, 2006 25th International Conference on Thermoelectrics 23 (2006).
[9] M. Wagner, G. Span, S. Holzer, and T. Grasser, Semiconductor Science and Technology **22**, S173 (2007).
[10] M. Wagner, G. Span, S. Holzert, and T. Grassert, Simulation 397 (2006).
[11] J.Y. Yang, T. Aizawa, a. Yamamoto, and T. Ohta, Materials Science and Engineering: B **85**, 34 (2001).
[12] P.L. Hagelstein and Y. Kucherov, Applied Physics Letters **81**, 559 (2002).
[13] N. Petermann, N. Stein, G. Schierning, R. Theissmann, B. Stoib, M.S. Brandt, C. Hecht, C. Schulz, and H. Wiggers, Journal of Physics D: Applied Physics **44**, 174034 (2011).
[14] A. Becker, G. Schierning, R. Theissmann, M. Meseth, and N. Benson, Journal of Applied Physics **111**, 054320 (2012).
[15] M. Meseth, P. Ziolkowski, G. Schierning, R. Theissmann, N. Petermann, H. Wiggers, N. Benson, and R. Schmechel, Scripta Materialia **67**, 265 (2012).
[16] S. Dongaonkar, J.D. Servaites, G.M. Ford, S. Loser, J. Moore, R.M. Gelfand, H. Mohseni, H.W. Hillhouse, R. Agrawal, M. a. Ratner, T.J. Marks, M.S. Lundstrom, and M. a. Alam, Journal of Applied Physics **108**, 124509 (2010).
[17] V. Kessler, D. Gautam, T. Hülser, M. Spree, R. Theissmann, M. Winterer, H. Wiggers, G. Schierning, and R. Schmechel, Advanced Engineering Materials n/a (2012).
[18] N. Stein, N. Petermann, R. Theissmann, G. Schierning, R. Schmechel, and H. Wiggers, (2011).

Mater. Res. Soc. Symp. Proc. Vol. 1543 © 2013 Materials Research Society
DOI: 10.1557/opl.2013.938

# Thermoelectric transport in topological insulator $Bi_2Te_2Se$ bulk crystals

Yang Xu[1,2], Helin Cao[1,2], Ireneusz Miotkowski[1], Yong P. Chen[1,2]
[1]Department of Physics, Purdue University, West Lafayette, IN 47907 U.S.A
[2]Birck Nanotechnology Center, Purdue University, West Lafayette, IN 47907 U.S.A
[4]School of Electrical and Computer Engineering, Purdue University, West Lafayette, IN 47907 U.S.A

## ABSTRACT

$Bi_2Te_2Se$ (BTS221) bulk crystals were recently discovered as an intrinsic 3D topological insulator. We have synthesized this material, and studied the transport properties of BTS221 from the thermoelectrics perspective. Temperature (T) dependent resistivity measurement indicates surface dominant transports in our sample at low T. We also report Seebeck measurement between 50K to room T.

## INTRODUCTION

There dimensional (3D) Topological insulator (TI), a new state of quantum matter, has attracted considerable attention mostly because of its unique topological protected surface states (TSS). On the surface of 3D TI materials, there exist non-trivial gapless surface states protected by time-reversal symmetry (TRS), whereas the bulk of TI resembles a normal band insulator. The TSS give rise to 2D helical Dirac fermions with linear energy momentum (E-k) dispersion. Due to the helical nature (resulting in spin-momentum locking) of the Dirac cone(s) on TSS, the topological protection reduces backscattering which requires spin flip (breaking the TRS), and is thus unlikely without magnetic impurities. Therefore, the TSS channels are largely immune to (nonmagnetic) structural imperfection. Because of the characteristic properties, TSS are promising for not only novel physics, but also nano-electronics[1]'[2]'[3]'[4]'[5]. Among other applications, TI materials are also expected to offer opportunities to enhance energy efficiency of thermoelectric (TE) devices[6][7]. We note that most of the current heavily studied TI materials, such as $Bi_2Se_3$ and $Bi_2Te_3$, are also widely-commercialized TE materials with high figure of merit (ZT~1, at ordinary temperatures, ~270 to 400 K). However, both $Bi_2Se_3$ and $Bi_2Te_3$ have substantial bulk conduction due to a significant amount of unintentional doping in the bulk. This has been one major challenge for the transport and device studies of TI, as a highly conducting bulk would "short-out" the TSS conduction and mask the transport features associated with TSS.

Recently, a new TI material, $Bi_2Te_2Se$ (BTS221) has been synthesized and demonstrated to have a dominant surface transport at low temperature (T) [8][9]. Here we report our preliminary thermal power (Seebeck coefficient) measurements on bulk BTS221 crystals. Since our BTS221 has been shown to be an intrinsic TI, with bulk insulating and surface conduction dominating at low T [10], studying the Seebeck coefficient of BTS221 at various temperatures (tuning from bulk dominated transport to surface dominated transport as T lowers) will likely provide much insights about thermoelectric transport of TSS and distinguish surface from bulk contributions.

## EXPERIMENT

We have synthesized high quality BTS221 crystals by Bridgman method from highly purified (99.9999%) elemental starting materials (Bi, Te and Se). Small pieces of single crystals were cleaved by razor blades, and then fabricated into quasi-Hall bar type devices by attaching contact leads with indium for various transport experiments. In order to perform Seebeck measurements in our cryostat, we built a home-made Seebeck stage. An optical image of the stage is shown in Fig. 1a, and Fig. 1b shows the schematic diagram. On our Seebeck stage, a piece of BTS221 crystal was suspended by two glass substrates. On each substrate, a platinum (Pt) stripe (thickness ~100 nm), serving as a thermometer, was deposited by e-beam evaporation. Temperature dependent four-terminal resistances of each Pt stripe were later measured individually. A pair of thermometers was used to probe temperature gradient $\Delta T$ cross the BTS221 crystal by monitoring the resistance change in the Pt stripes. Finally, a film heater was attached to one end of the crystal across which a temperature gradient can be generated by passing electrical current through the heater. After tuning on the heater, we measure voltage difference caused by $\Delta T$, and then calculate S by $\Delta V/\Delta T$. Here we present a representative result measured in a BTS221 crystal (L×W×T ~ 10mm×3mm×200μm).

## RESULTS AND DISCUSSION

Figure 1c shows a 4-terminal longitudinal resistance as a function of T from 1.8K to 220K by slowly warming up the sample. The resistivity increases as temperature decreases, indicating an insulating behavior. Our data can be fitted to Arrhenius law ($R_{xx} \sim e^{\Delta/kT}$, where k is the Boltzmann constant, and $\Delta$ is an activation energy gap used as the only fitting parameter) very well with $\Delta$ ~35meV from 88 K to 220 K as shown in the inset of Fig. 1c. Below 20K, the resistivity begins to saturate. It suggests a surface conduction dominated region. At base temperature, the resistivity reaches ~6 Ω*cm which is comparable to the largest bulk resistivity values have ever been reported in this material[9].

Seebeck coefficient of the BTS221 crystal measured in a mediate T range (between 50 K and 290 K) was shown in Fig. 1d. The positive S indicates a p-type carrier, consistent with the Hall measurement in this sample (data not shown here). As T increases up to room T, S slightly increases. The values of our measured S are found to be comparable to previous measurements [11]. We believe the Seebeck coefficient measured in the mediate T range still involve substantial contribution from the bulk. Interestingly, the value of S at 50 K is notable, and only ~30% lower than that at room T. Future measurements at lower T may be helpful to resolve the surface contribution.

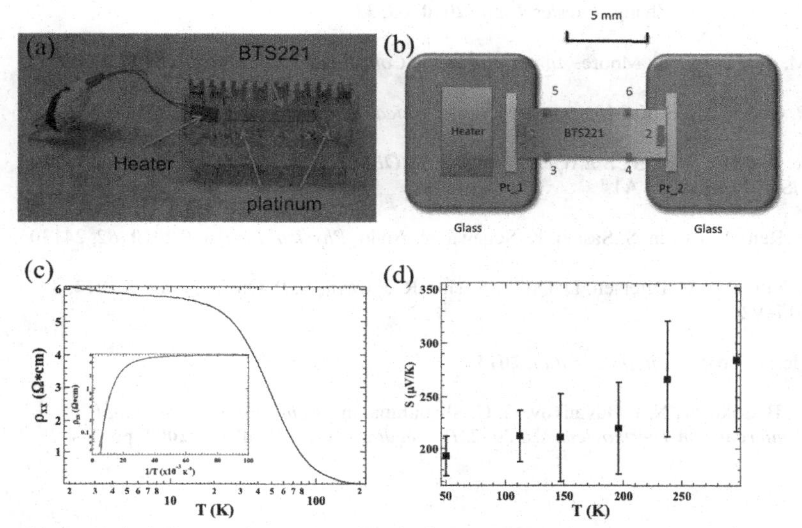


Fig.1. (a) An optical image of one $Bi_2Te_2Se$ single crystal sample mounted on a home-made Seebeck measurement stage. (b) Schematic diagram. (c) Four-terminal $R_{xx}$ (also resistivity) as a function of temperature. An excitation energy ~ 35 meV was extracted by fitting the data to Arrhenius law (as shown in the inset with blue line showing the experimental data and red line the fitting curve). (d) Seebeck coefficient measured in the sample between 50 K and 290 K.

## CONCLUSIONS

We have synthesized BT221 crystal. The bulk of the sample is found to be very insulating with a resistivity of ~ 6 Ω*cm at 1.8 K. A home-made Seebeck measurement stage was used to measure the Seebeck coefficient in cryostat. The Seebeck coefficient of our crystal was found to be p-type above 50 K, and can be useful for thermoelectric applications.

## ACKNOWLEDGMENTS

We acknowledge support from DARPA MESO program (Grant N66001-11-1-4107).

## REFERENCES

[1] J. E. Moore, *Nature* **2010**, *464*, 194–198.

[2] M. Hasan, C. Kane, *Reviews of Modern Physics* **2010**, *82*, 3045–3067.

[3] H. C. Manoharan, *Nature nanotechnology* **2010**, *5*, 477–479.

[4] X.-L. Qi, S.-C. Zhang, *Physics Today* **2010**, *63*, 33.

[5] M. Z. Hasan, J. E. Moore, *Annual Review of Condensed Matter Physics* **2011**, *2*, 55–78.

[6] P. Ghaemi, R. S. K. Mong, J. E. Moore, *Physical Review Letters* **2010**, *105*, 166603.

[7] Y. P. Chen, *SURFACE EXCITONIC THERMOELECTRIC DEVICES*, **2012**, U.S. Patent US 2012/0138115 A1.

[8] Z. Ren, A. Taskin, S. Sasaki, K. Segawa, Y. Ando, *Physical Review B* **2010**, *82*, 241306.

[9] J. Xiong, A. C. Petersen, D. Qu, Y. S. Hor, R. J. Cava, N. P. Ong, *Physica E* **2012**, *44*, 917–920.

[10] Helin Cao et al., *in preparation* **2013**.

[11] O. B. Sokolov, N. I. Duvankov, G. G. Shabunina, in *Twenty-First International Conference on Thermoelectrics, 2002. Proceedings ICT '02.*, IEEE, **2002**, pp. 1–4.

Mater. Res. Soc. Symp. Proc. Vol. 1543 © 2013 Materials Research Society
DOI: 10.1557/opl.2013.605

# Large Electrocaloric Effect from Electrical Field Induced Orientational Order-Disorder Transition in Nematic Liquid Crystals Possessing Large Dielectric Anisotropy

Xiao-Shi Qian[1], S. G. Lu[2], Xinyu Li[1], Haiming Gu[1], L-C Chien[3], Q. M. Zhang[1,2]
[1]Department of Electrical Engineering and [2]Materials Research Institute
The Pennsylvania State University, University Park, PA 16802, USA
[3]Liquid Crystal Institute and Department of Chemical Physics,
Kent State University, Kent, OH 44242, USA

## ABSTRACT

Large electrocaloric (EC) effects in ferroelectric polymers and in ferroelectric ceramics have attracted great attention for new refrigeration development which is more environmental friendly and more efficient and thus could be an alternative to the existing vapor-compression refrigerators which consume large energy and release large amount of green house gas. However in the past, all EC effects investigations have been focused on solid state dielectrics. It is interesting to ask whether a large EC effect can also be realized in dielectric fluids. A dielectric fluid with large EC effect could lead to new design of cooling devices with simpler structures than these based on solid state EC materials, for example, they can be utilized as both the refrigerant and heat exchange fluid. Here we present that a large EC effect can be realized in the liquid crystal (LC) 5CB near it's nematic-isotropic (N-I) phase transition. The LC 5CB possesses a large dielectric anisotropy which can induce large polarization change from the isotropic phase to the nematic phase near the N-I transition. An isothermal entropy change of more than 23 $Jkg^{-1}K^{-1}$ was observed near 39 °C that is just above the N-I transition.

## INTRODUCTION

The electrocaloric (EC) effect is a reversible temperature and/or entropy change of an insulating polar material under applying and removing of an electric field E [1-5]. Dielectrics with a large EC effect are attractive for developing alternatives for conventional mechanical vapor compression cycle (MVCC) based air-conditioning and refrigeration which use strong greenhouse gases as the refrigerant, e.g., operating air-conditioning only with electrical power is of great benefit to electrical vehicles considering an extra-burden of the mechanical compressor required for MVCC. The recent findings of giant EC effects in ferroelectric polymers and in ferroelectric ceramics have attracted great interest for designing new cooling devices which are environmental friendly, compact, and more efficient [3-18].

In all refrigeration and air-conditioning, two key components are required. First, it is required for a refrigerant whose entropy depends on some properties other than temperature, e.g., pressure or magnetic field. Second, entropy needs to be transported from one temperature level to another temperature level in a reversible and cyclic manner in order to get high efficiency. In EC materials, insulating dielectric materials, whose entropy can be changed by external electric fields, play a role as a refrigerant. In order to transport entropy, various methods are developed. Among them, heat exchange fluid is widely used in heat transfer and passive cooling and is commercialized to different products to meet various demands. Also, it is demonstrated that heat exchange fluid can be effectively used to transport entropy in cooling devices that utilize magnetic alloys with giant magnetocaloric effect (MCE) as refrigerant [19, 20]. It is interesting to ask whether dielectric fluids could perform EC effect and be able to use as active cooling refrigerant. A fluid EC material would be more attractive for its capabilities as both the refrigerant and heat exchange fluid, leading to diverse designing of new cooling devices.

To electrically induce larger dipolar entropy change, it is better for an EC material to enable high dipole density and high electrical breakdown strength [1-9]. In addition, a dipolar disordered state is much easier tuned by electric field to an ordered state by electric field near and above dipolar order-disorder transition temperature [1-4]. Liquid crystals (LCs), which have been widely used in optic displays and optic modulators, are dipolar liquids in which the molecular orientation and consequently dipolar states can be easily controlled by external electric signals [21-23]. Hence LCs show great opportunities to realize large EC response. Applying electrical fields may induce a transition from an isotropic (I) phase to a nematic (N) or smectic (S) phase when near an N-I or S-I transition, thus may lead to a large EC response [21, 22], especially near transition temperature. In this paper, we investigate the EC response in LC 5CB (4-n-pentyl-4'-cyanobiphenyl) near its N-I transition temperature. 5CB is chosen for its N-I transition temperature (~ 35 °C) that is close to room temperature. Moreover, 5CB mesogens show large dielectric anisotropy. Its dielectric constant parallel to the director ($\varepsilon_{\parallel}$>17.5) is much larger than that perpendicular to the director (~8), which provides a strong orientation response for electric modulation [24-26].

## EXPERIMENT

The molecular orientations of LCs can be strongly influenced by the surface conditions of substrates with which LC molecules are in contact and consequently the entropy of the LCs [27-29]. In order to more generally study the effect, two groups of LC cells, i.e., homogeneously aligned cells (HA-Cells) and no-aligned cells (NA-Cells), are purchased from Instec for HA-Cells (parts No. SA025A032uG180) and NA-Cells (SA100A040uNOPI). In the HA-Cells, the rodlike LC molecules are aligned preferentially along the surface of the substrates. In this study, the HA-Cells have a thickness of 3.2 μm. In the cell without surface alignment layer, the cell is made by two transparent ITO electrodes coated glass substrates separated by glass fiber spacers with a thickness d = 4 μm (NA-Cells).

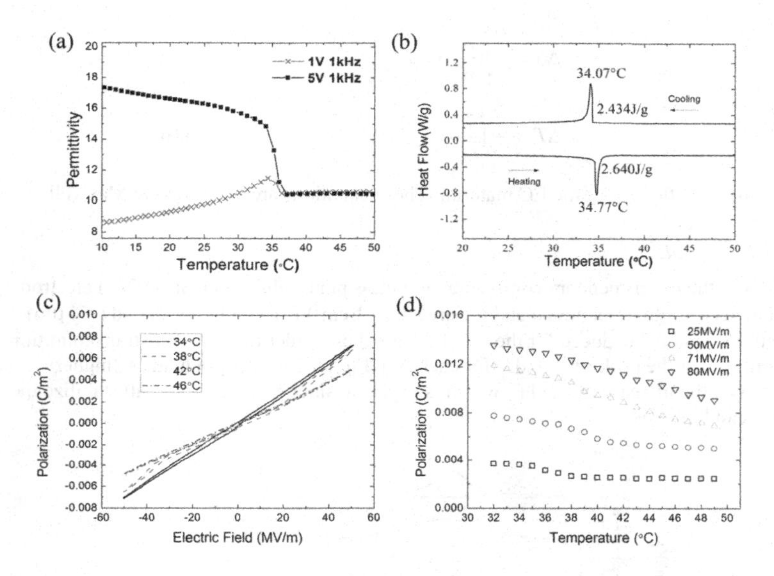


**Figure 1**. (a) Temperature-dependent low field dielectric properties of 5CB in HA-Cells. (b)DSC data of 5CB liquid crystal measured during the heat and cooling runs. The DSC peak corresponding to the N-I transition is at 34.7 °C in heating and 34.07 °C in cooling processes. (c) D-E loops of 5CB versus temperature near N-I transition temperature. (d) Polarization change with temperature under different electric fields.

The dielectric properties as a function of temperature were characterized using a multi-frequency LCR Meter (HP 4284A) equipped with a temperature controlled chamber. The electric displacements vs. electric fields (D-E) loops sweeping different temperatures were measured using a Sawyer-Tower circuit with a temperature chamber. Figure 1(a) shows the dielectric data versus temperature under voltages below and above threshold voltage. Large dielectric anisotropy enables large orientational response to electric fields. Presented in Figure 1(c) presents the polarization responses (electric displacement-electric field loops, D-E loops) of 5CB in HA-Cells, measured under 50 $MVm^{-1}$ and 100 Hz AC field at temperatures near N-I transition. Detailed polarization change with temperature under different external fields is shown in Figure 1(d). The data reveal that there is very little hysteresis in the D-E loops and the polarization decrease corresponding to the increasing temperature. For the LC 5CB studied here, it has a weak first order N-I transition around 34.5 °C which has a small thermal hysteresis of 0.7 °C of the N-I transition between the heating and cooling runs as shown in Figure 1(b). Hence the Maxwell relation can be used to deduce EC effect at temperatures above the critical point. Reversible isothermal entropy change ΔS and adiabatic temperature change ΔT for a dielectric film as the electric field changes from $E_1$ to $E_2$ can be deduced from the pyroelectric coefficient $\partial D/\partial T$ as a function of electric field (Figure 2), i.e. [1-5],

$$\Delta S = -\int_{E1}^{E2} (\frac{\partial D}{\partial T})_E dE \qquad (1a)$$

and

$$\Delta T = -\int_{E1}^{E2} \frac{T}{c_E} (\frac{\partial D}{\partial T})_E dE \qquad (1b)$$

where $C_E$ is the specific heat of the EC material. These equations are based on the Maxwell equation $(\frac{\partial D}{\partial T})_E = (\frac{\partial S}{\partial E})_T$.

The LCs in the experiment are confined by the glass plate cells which have fixed electrode area A and dielectric film thickness d and the area A (~ 1 $cm^2$) is much larger than d (= 4 μm). Consequently ΔS and ΔT deduced for the LCs here are these under the constant strain condition. It is also noticed that the peak of $\partial D/\partial T$ for each E in Figure 2 moves gradually to higher temperature with the increase of field E, which is expected since a higher field will stabilize the nematic phase to higher temperatures.

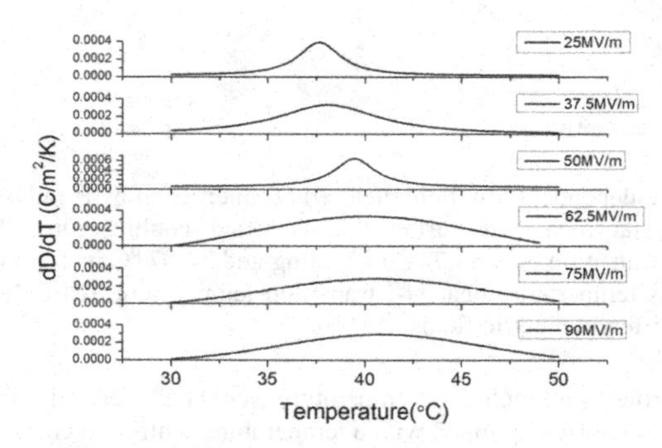


**Figure 2**. Pyroelectric coefficient $\partial D/\partial T$ as a function of electric field.

The isothermal entropy changes ΔS deduced as a function of temperatures are presented in Figure 3(a) and Figure 3(b) for LCs measured in NA-Cells and HA-Cells, respectively. The data reveal that EC response peaks near N-I transition which has ΔS= 23.6 $Jkg^{-1}K^{-1}$ under applied electric field of 90 $MVm^{-1}$ for the 5CB in the NA-Cells. The EC effect observed here is nearly the same as that observed in the polar-fluoropolymers under the same electric field level, which also exhibit very large EC effect near room temperature [4, 8]. As shown earlier, increasing applied field will stabilize the nematic phase to higher temperatures. Hence, the EC response peak will shift progressively towards higher temperature as observed experimentally in Figure 3(a, b).

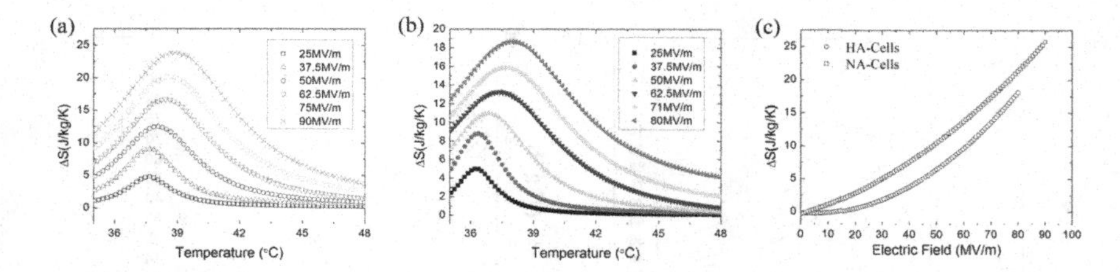


**Figure 3**. The electrocaloric effect deduced from the Maxwell relations for (a) isothermal entropy change ΔS of 5CB LCs in NA-Cells, and (b) ΔS of 5CB LCs in homogeneously aligned LC cells (HA-Cells). Data points are shown and solid curves are drawn to guide eyes. (c) Comparison of ΔS as a function of electric field E of 5CB at 39 °C in NA-Cells (crosses) and in HA-Cells (squares). Data pointes are shown and solid curves are drawn to guide eyes.

Figure 3(c) compares ΔS as a function of applied field amplitude for the two groups of LC cells at 39 °C which is slightly above the N-I transition. It is interesting to find that the LCs in NA-Cells show higher ΔS than that with those in HA-Cells, since surface alignment induces partial ordering of molecular alignment in the LC films and reduces the dipolar entropy in E=0 state.

## DISCUSSION

The phenomenological Landau-de Gennes formulation (LG) can also be used to estimate entropy change in LCs [21]. The isothermal entropy change ΔS of a nematic liquid crystal at temperatures above the N-I transition critical point is related to the change of the order parameter Q ($0 \leq Q \leq 1$) (we use Q since S is already occupied by entropy) as (Figure 4) [21, 27]

$$\Delta S = (a/2)\,(Q_1^2 - Q_2^2) \qquad (2)$$

where $Q_1$ and $Q_2$ are the order parameter at electric fields $E_1$ and $E_2$, and $a$ is a constant. It is obvious that Q=1 corresponds to a perfect LC molecular alignment which may be achieved under a very high electric field E while Q=0 is associated to molecular random orientation as in the isotropic phase. From the $a$ value reported in the literature for 5CB [25, 26], $a = 1.3 \times 10^5$ $JK^{-1}m^{-3}$ and the density $\rho = 1$ $gcm^{-3}$, an very large ΔS =65 $Jkg^{-1}K^{-1}$ can be deduced when Q is changed from $Q_1$=0 to $Q_2$=1. For the study here $Q_2$=0.65 is induced under 90 MV/m which yields a ΔS = 27.5 J $kg^{-1}K^{-1}$. The agreement between the experimental data and LG phenomenological theory estimation is quite good considering the uncertainty of the coefficient $a$ and the possible substrate effects in the thin LC cells. These results indicate that giant EC response can be obtained from electrical field induced orientational order-disorder transition in LCs near N-I and S-I transitions.

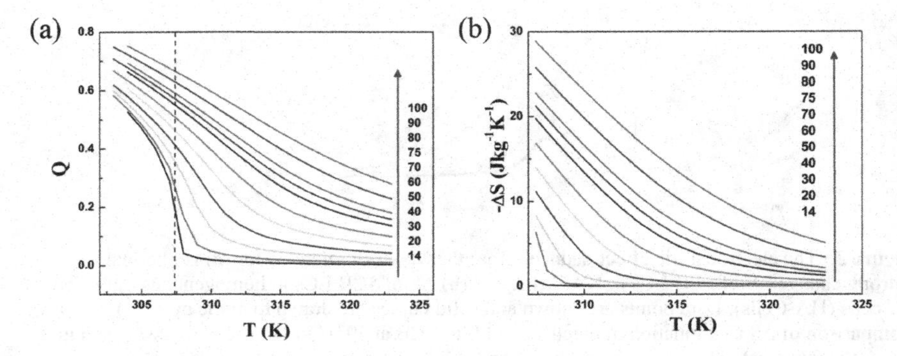


**Figure 4**. (a) Order parameter Q as a function of temperature under different electric fields for 5CB. Parameters used here are from ref. [26]. The electric field changes from 14 $MVm^{-1}$ (critical field) to 100 MV/m. (b) Isothermal entropy change as a function of temperature under different electric fields for 5CB. Parameters used here are from Ref. [26] and Figure 4(a). The electric field changes from 14 $MVm^{-1}$ (critical electric field) to 100 $MVm^{-1}$.

Other than 5CB, there are a broad range of LCs available in which the N-I and S-I transition temperatures cover a broad temperature window [30]. In addition, it is easy to tune the LC phase transition temperature by, e.g., mixing multiple LCs together with different composition [31]. For cooling devices to be operated over a broad temperature range, LCs with large EC effect at different temperature regions may be cascaded to cover the required temperature window. This is commonly used in cooling devices based on the magnetic alloys possessing giant magnetocaloric effect (MCE) which also shows a relatively narrow working temperature range that near the ferromagnetic-paramagnetic transition temperature [32].

## CONCLUSION

We investigated the EC effect in a LC material that shows large dielectric anisotropy near its N-I transition temperature. The large dielectric anisotropy in the liquid crystal 5CB facilitates the electric field induced large polarization change and thus a large EC response, *i.e.*, an isothermal entropy change of more than $23Jkg^{-1}K^{-1}$, was observed near 39 °C, right above the N-I transition temperature region. One can expect a wide working temperature for LC-based cascaded cooling system considering broad availability of different LCs and their mixtures that have different phase transition temperatures cascaded to cover a wide window. Moreover, fluidic EC materials could lead to various and more flexible design of EC-based cooling devices since it can flow in channels in various shapes.

## ACKNOWLEDGMENTS

Xiao-Shi Qian, Xinyu Li, and Q. M. Zhang were supported by the U.S. Department of Energy Division of Materials Sciences through grant No. DE-FG02-07ER46410. S. G. Lu and Haiming Gu were supported by Army Research Office under grant W911NF-11-1-0534. We thank Antal Jakli and I-C Khoo for stimulating discussions.

**REFERENCES**


[1] Lines, M. & Glass, A. *Principles and Applications of Ferroelectrics and Related Materials.* (Larendon Press, Oxford, 1977).
[2] E. Fatuzzo and W. J. Merz, *Ferroelectricity* (North-Holland, Amsterdam, 1967).
[3] A. S. Mischenko, Q. Zhang, J. F. Scott, R. W. Whatmore, and N. D. Mathur, Science **311**, 1270 (2006).
[4] B. Neese, B. Chu, S. G. Lu, Y. Wang, E. Furman, and Q. M. Zhang, Science **321**, 821 (2008).
[5] S. G. Lu and Q. M. Zhang, Adv. Mater. **21**, 1983 (2009).
[6] T. M. Correia et al. Appl. Phys. Lett. **95**, 182904 (2009).
[7] S. G. Lu, B. Rožič, Q. M. Zhang, Z. Kutnjak, X. Li, E. Furman, Lee J. Gorny, M. R. Lin, B. Malič, M. Kosec, R. Blinc, R. Pirc, Appl. Phys. Lett. **97**, 162904 (2010).
[8] X. Li, X. S. Qian, S. G. Lu, J. P. Cheng, Z. Fang, Q. M. Zhang, Appl. Phys. Lett. **99**, 052907 (2011).
[9] Z. K. Liu, Xinyu Li, and Q. M. Zhang, Appl. Phys. Lett. **101**, 082904, (2012).
[10] L. Liu, Y. Liu, J. Leng, Appl. Phys. Lett. **99**, 181908, (2011).
[11] G. Akcay, S. P. Alpay, J. Mantese, G. A. Jr. Rossetti, Appl. Phys. Lett. **90**, 252909, (2007).
[12] X. Li, X.-S Qian, H. Gu, X.-Z Chen, S. G. Lu, Minren Lin, Fred Bateman, and Q. M. Zhang. Appl. Phys. Lett. **101**, 132903 (2012).
[13] X.-Z. Chen, X.-S Qian, X. Li, S.G. Lu, H. Gu, M. Lin, Q.-D. Shen, and Q. M. Zhang. Appl. Phys. Lett. **100**. 222902, (2012).
[14] S. Kar-Narayan, S. Crossley, X. Moya, V. Kovacova, J. Abergel, A. Bontempi, N. Baier, E. Defay and N. D. Mathur. Appl. Phys. Lett. **102**, 032903 (2013).
[15] M. C. Rose and R. E. Cohen. Phys. Rev. Lett. **109**, 187604 (2012).
[16] Y. Bai, G.-P. Zheng, K. Ding, L. Qiao, S.-Q. Shi, and D. Guo. J. Appl. Phys. **110**, 094103 (2011).
[17] Y. Jia and Y. S. Ju. Appl. Phys. Lett. **100**, 242901 (2012).
[18] S. Kar-Narayan, N. D.Mathur. Ferroelectrics (0015-0193), **433** (1), 107, (2012).
[19] Y. V. Sinyashii, *Chem. & Petro. Eng.* **31**, 501 (1995).
[20] Y. V. Sinyavsky, G. E. Lugansky, N. D. Pashkov, *Cryogenics*, **32**, 28, (1992).
[21] P. G. De Gennes & J. Prost, *The Physics of Liquid Crystals* (2nd edit. Clarendon Press, Oxford, 1995).
[22] L. M. Blinov and V. G. Chigrinov, *Electrooptic Effect in Liquid Crystal Materials* (Springer, NY, 1996).
[23] W. Lehmann et al., Nature **410**, 447 (2001).
[24] K. Abe, A. USami, K. Ishida, Y. Fukushima, and T. Shigenari, J. Korean Phys. Soc. **46**, 220 (2005).
[25] H. J. Coles, Mol. Cryst. Liq. Cryst. Lett. **49**, 67 (1978).
[26] I. Lelidis and G. Durand, Phys. Rev. **E48**, 3822 (1993).

[27] G. P. Crawford, R. Ondris-Crawford, S. Zumer, and J. W. Doane, Phys. Rev. Lett. **70**, 1838 (1993).
[28] J. Bechhoefer *et al.*, Phys. Rev. Lett. **64**, 1911 (1990).
[29] J. S. Patel and H. Yokoyama, Nature **362**, 525 (1993).
[30] D. Demus, J. Goodby, G. W. Gray, H. –W. Spiess, V. Vill, *Handbook of Liquid Crystals* (Wiley-VCH, Weinheim 1998).
[31] D.-K Yang, Y. Yin, and H. Liu, Liq. Cryst. **34**, 605 (2007).
[32] A. M. Tishin and Y. I. Spichkin, *The Magnetocaloric Effect and Its Applications* (IOP Publishing Ltd, Bristol 2003).

## Modeling/Theory

**Mater. Res. Soc. Symp. Proc. Vol. 1543 © 2013 Materials Research Society**
**DOI: 10.1557/opl.2013.969**

# First-principles investigations on the thermoelectric properties of $Bi_2Te_3$ doped with Se

Liwen F. Wan and Scott P. Beckman
Department of Material Science and Engineering, Iowa State University, Ames, IA 50010, U.S.A.

## ABSTRACT

In this work, the thermoelectric properties of Se-doped $Bi_2Te_3$ are examined using first-principles density functional theory and semi-classical Boltzmann transport theory. Placing a single Se atom on the *3a* Wyckoff position lowers the unit cell energy by approximately 3.6 eV, compared to the *6c* Te position. The electronic structure of $Bi_2Te_3$ has minor changes upon Se doping. At carrier concentration of $10^{19}$ $cm^{-3}$, the optimal thermopower, $S$, is obtained as 207 and 220 μV/K for n-type and p-type doping, respectively. Unlike the thermopower, the power factor, $S^2\sigma/\tau$, is highly anisotropic for the in-plane and cross-plane conduction. At carrier concentrations of $10^{19}$ $cm^{-3}$, the best power factor is predicted to be around 1.05 and $1.4\times10^{11}$ $W/m{\cdot}s{\cdot}K^2$ for n-type and p-type doping, respectively.

## INTRODUCTION

$Bi_2Te_3$ is a well-known thermoelectric (TE) material that can be used near room temperature [1-3]. The potential energy conversion efficiency of a typical TE material is measured by the dimensionless figure of merit, $ZT = S^2\sigma T/\kappa$, where $S$ is the Seebeck coefficient (also referred to as thermopower), $T$ is the operating temperature, $\sigma$ and $\kappa$ are the electrical and thermal conductivity, respectively. The thermal conductivity is further divided into the lattice contribution, $\kappa_l$, and the electronic contribution, $\kappa_e$, where $\kappa_e$ scales almost linearly with $\sigma T$. In practice, it is extremely challenging to obtain a large $ZT$ value because the material properties are interrelated and tend to counter one another. For example, a material with a high Seebeck coefficient usually has fewer free carriers, which means a low electrical conductivity is often exhibited. On the other hand, increasing the electrical conductivity also increases the electronic contribution to the thermal conductivity. Ultimately, the pragmatic approach is to search for heavily doped semiconductors that also have a large $\sigma/\kappa$ ratio.

The $Bi_2Te_3$ crystal family has been intensively studied, and $ZT$ values greater than 1 are reported in the literature [3-5]. Many scientific attempts have been made to improve their TE performance by controlling the thermal conductivity [6-13]. At the atomic scale, it is shown that the $\kappa_l$ can be greatly reduced by introducing point defects, which help to scatter phonons [7,8]. In addition, by forming long-ranged periodic structures, such as superlattices, the TE performance may also be enhanced [11,12].

Here we examine the effect of Se doping on the TE properties of $Bi_2Te_3$, by introducing a single Se atom to the hexagonal $Bi_2Te_3$ unit cell. The electronic structure of Se-doped $Bi_2Te_3$ is determined from first-principles density functional theory (DFT) [14,15]. Combining DFT with the semi-classical Boltzmann transport theory [16,17], the TE behavior is discussed in relation to the free carrier concentration and temperature.

## METHOD

Within DFT [14], the Kohn-Sham eigenfunctions are represented using the linearized augmented plane wave method (LAPW) [18] as implemented in the WIEN2k software [19]. The values of atomic sphere radii (RMT) are chosen as 2.5 a.u. for Bi, 2.3 a.u. for Te and Se. This basis set is well converged with $R_{min}K_{max} = 7.0$, where $R_{min}$ is the minimum sphere radius and $K_{max}$ is the maximum cutoff for interstitial plane waves. An all-electron potential is used for each atomic species, and a fully relativistic approach is used for the core states. The spin-orbit coupling effects are included for the valence states via a second variational approach. The exchange-correlation energy is approximated using the PBE-GGA functional [20]. The ground state of the system is solved by sampling 1728 *k* points on a shifted uniform distributed grid [21]. For the transport properties, a much denser k-point mesh is used, 27648, to assure that the conductivity tensor is fully converged. The TE properties are determined using BoltzTraP [17].

## RESULTS AND DISCUSSIONS

The crystal structure of $Bi_2Te_3$ belongs to the space group R-3m. The 5-atom rhombohedral unit cell is shown in Fig. 1(a). It is also convenient to represent the system using a hexagonal representation. As shown in Fig. 1(b), the stacking sequence of $Bi_2Te_3$ along the hexagonal c-axis is Te(1) – Bi – Te(2) – Bi – Te(1). Here, the lattice parameters are taken from the experimental literature: a = b = 4.3835 Å and c = 30.487 Å [22]. It is found that the energy to substitute Se to the Te(2) site is 3.6 eV lower than to replace the Te(1) atom with Se.

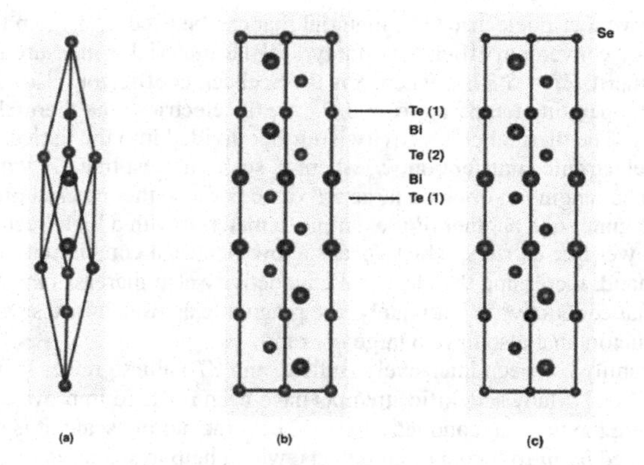


Figure 1: The crystal structure of $Bi_2Te_3$. (a) Rhombohedral representation of $Bi_2Te_3$ unit cell. (b) Hexagonal representation of $Bi_2Te_3$ unit cell. (c) Hexagonal representation of $Bi_2Te_3$ with Se doped on the top layer.

Experimentally, it is observed that coating the $Bi_2Te_3$ hexagonal cell with a single layer of Se can significantly improve the TE performance of the system [23]. In this study, the top layer of Te in the $Bi_2Te_3$ hexagonal unit cell is replaced by Se, as shown in Fig. 1(c). This is the low energy Te(2) site. In this initial set of calculations, no structural relaxations are performed because these will yield only minor changes to the electronic structure.

In Fig. 2(a), the Kohn-Sham energy states are plotted for Se-doped $Bi_2Te_3$ in reciprocal space. Integrating the electronic states in the first Brillouin zone gives the total density of states, as presented in Fig. 2(b). Compared to bulk $Bi_2Te_3$, there is no significant change to the electronic structure. However, the GGA band gap is slightly reduced from 170 meV to 140 meV.

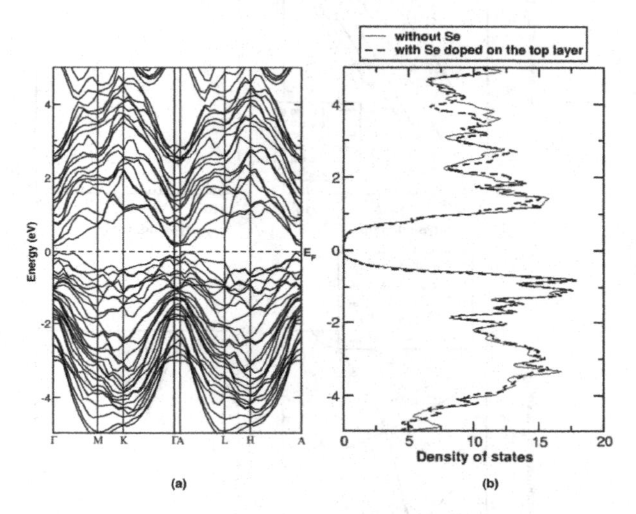


Figure 2: The electronic structure of Se-doped $Bi_2Te_3$. (a) Band structure plotted along the high-symmetry path in the first Brillouin zone. (b) Total density of states. The energy zero is set at the Fermi level.

Within the constant relaxation time approximation, the transport coefficients are calculated by analyzing the band energies of the system. Because of the observed narrow band gap, bipolar conduction is expected for the $Bi_2Te_3$ related structures. In Fig. 3, the thermopower, $S$, is presented as a function of doping level for temperatures ranging from 100 K to 350 K. A "roll-over" phenomenon is identified for the thermopower at low doping level or high temperature as we expected from the double-sign conduction. The optimal thermopower can be obtained for a specific temperature or doping level. For example, at carrier concentrations of $10^{19}$ $cm^{-3}$, the highest $S$ values for Se-doped $Bi_2Te_3$ are around 207 and 220 μV/K for electron- and hole-type conduction, respectively. It is clear from Fig. 3(c) and (d) that the crystal structures examined here are indeed only suitable for near room temperature applications, where $S$ has a maxima for both n-type and p-type doping. Although this peak can shift to a higher temperature at extremely high doping levels, the magnitude of the TE response will be dramatically reduced.

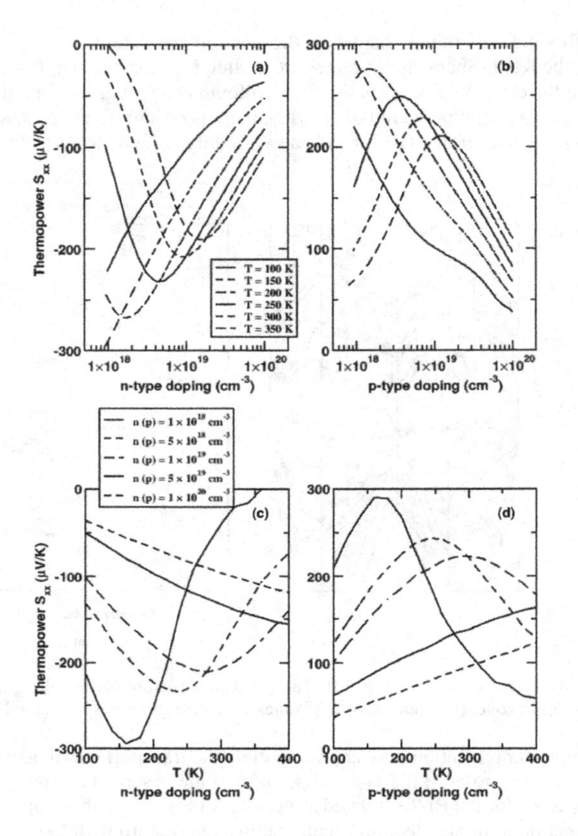


Figure 3: The Seebeck coefficient $S$ for Se-doped $Bi_2Te_3$. (a) – (b) $S$ plotted as functions of doping level at different temperature. (c) – (d) $S$ plotted as functions of temperature at different doping level.

In addition to $S$, the power factor, $PF = S^2\sigma$, is also calculated and plotted in Fig. 4. Unlike $S$, which is expected to be isotropic, the power factor shows a significant difference for the in-plane and cross-plane conduction. A higher $PF$ value is always found for the in-plane conduction ($xx$), although for the p-type doping, the difference of $PF$ between the in-plane and cross-plane conduction are much smaller. At room temperature (300 K), the maximum $PF$ is approximately $1.1 \times 10^{11}$ W/m·s·K$^2$ at electron concentrations of $2 \times 10^{19}$ cm$^{-3}$. The $PF$ is $1.65 \times 10^{11}$ W/m·s·K$^2$ for hole concentrations of $4 \times 10^{19}$ cm$^{-3}$. Above room temperature, there is still a substantial increase of the $PF$ value; however, the peak is shifted to a much higher doping level, which will be challenging to achieve experimentally.

It should be noted here that the $PF$ value in Fig. 4 is scaled by the relaxation time, $\tau$. Theoretically, it is non-trivial to obtain this because there are numerous scattering mechanisms occurring simultaneously. Using the experimental determined $\tau$ for bulk $Bi_2Te_3$, which is $2.2 \times 10^{-14}$ s at 300 K, and assuming that this also applies for the Se-doped crystals, the maximum *PF* value for Se-doped $Bi_2Te_3$ is approximated to be 24 and 36 $\mu W/cm \cdot K^2$ for n-type and p-type doping, respectively. The power factors predicted here are slightly lower than the values obtained for bulk $Bi_2Te_3$ [24], which suggests that the enhanced TE performance due to Se-doping is likely due to changes to the thermal conductivity, not an increase of power factor.

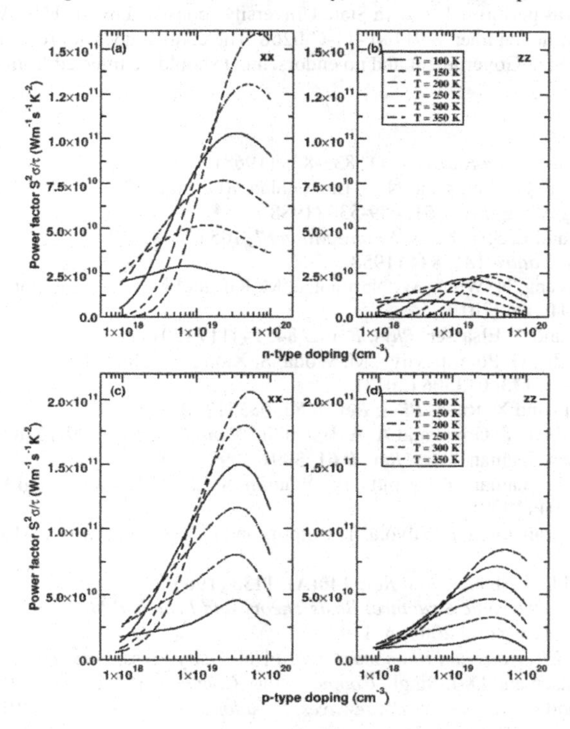


Figure 4: The power factor, *PF*, plotted as a function of doping level. (a) – (b) In-plane (*xx*) and cross-plane (*zz*) results for n-type doping. (c) – (d) Results for p-type doping.

## CONCLUSIONS

In this work, the impact of Se-doping on the TE properties of $Bi_2Te_3$ is studied. By computing the energy difference for placing Se on different Te sites, it is found that the Te(2) site is favored by approximately 3.6 eV. Incorporating a layer of Se into $Bi_2Te_3$ has negligible impact on the shape of the band structure; however, the band gap energy is reduced from 170 meV to 140 meV. The thermopower of Se-doped $Bi_2Te_3$ is greater than 200 $\mu V/K$ at n (p) = $10^{19}$ $cm^{-3}$. The power factor is strongly anisotropic comparing the in-plane and cross-plane conduction due to the differences in bonding. A higher power factor is always obtained for the in-plane conduction, with the maximum value of 1.05 and $1.4\times10^{11}$ W/m·s·K$^2$ for n-type and p-type doping at carrier concentrations of $10^{19}$ cm$^{-3}$.

## ACKNOWLEDGMENTS

This work was performed at Iowa State University sponsored by the U.S. Army Research Office under contract number W911NF-11-C-0268. The content does not reflect the position or the policy of the U.S. Government and no endorsement should be inferred from this sponsorship.

## REFERENCES

1. F. D. Rosi, *Solid-State Electron.* **11,** 833-848 (1968).
2. F. D. Rosi, E. F. Hockings and N. E. Lindenblad, *RCA Rev.* **22**, 82-121 (1961).
3. C. Wood, *Rep. Prog. Phys.* **51**, 459-539 (1988).
4. G. J. Snyder and E. S. Toberer, *Nature Mater.* **7**, 105 (2008).
5. D. A. Wright, *Nature* **181**, 834 (1958).
6. C. Wan, Y. Wang, N. Wang, W. Norimatsu, M. Kusunoki, and K. Koumoto, *Sci. Technol. Adv. Mater.* **11**, 044306 (2010).
7. A. Hashibon and C. Elsässer, *Phys. Rev. B* **84**, 144117 (2011).
8. K. Termentzidis, O. Pokropyvnyy, M. Woda, S. Xiong, Y. Chumakov, P. Cortona and S. Volz, *J. Appl. Phys.* **113**, 013506 (2013).
9. B. Qiu, L. Sun and X. Ruan, *Phys. Rev. B* **83**, 035312 (2011).
10. D. Teweldebrhan, V. Goyal and A. A. Balandin, *Nano Lett.* **10**, 1209-1218 (2010).
11. R. Venkatasubramanian, *Phys. Rev. B* **61**, 3091-3097 (2000).
12. R. Venkatasubramanian, T. Colpitts, B. O'Quinn, S. Liu, N. El-Masry and M. Lamvik, *Appl. Phys. Lett.* **75**, 1104 (1999).
13. R. Venkatasubramanian, E. Silvola, T. Colpitts and B. O'Quinn, *Nature* **413**, 597-602 (2001).
14. W. Kohn and L. J. Sham, *Phys. Rev.* **140(A)**, 1133 (1965).
15. R. M. Martin, *Electronic Structure: Basis Theory and Practical Methods* (Cambridge University Press, New York, 2004) p. 119.
16. J. M. Ziman, *Electrons and Phonons* (Oxford University Press, New York, 2001) p. 257.
17. G. K. H. Madsen and D. J. Singh, *Comput. Phys. Commun.* **175**, 67-71 (2006).
18. D. J. Singh and L. Nordström, *Planewaves, Pseudopotentials and the LAPW Method*, 2nd ed. (Springer, New York, 2006) p. 43.
19. P. Blaha, K. Schwarz, G. Madsen, D. Kvasnicka, and J. Luitz, *WIEN2k, An Augmented Plane Wave+Local Orbitals Program for Calculating Crystal Properties* (K. Schwarz Technical University, Wien, Austria, 2001) p. 1.
20. J. P. Perdew, K. Burke, and M. Ernzerhof, *Phys. Rev. Lett.* **77**, 3865 (1996).
21. H. J. Monkhorst and J. D. Pack, *Phys. Rev. B* **13**, 5188 (1976).
22. R. W. G. Wyckoff, *Crystal Structures* (John Wiley and sons, New York, 1964) p. 10.
23. S. Wang, G. Tan, W. Xie, G. Zheng, H. Li, J. Yang and X. Tang, *J. Mater. Chem.* **22**, 20943 (2012).
24. N. F. Hinsche, B. Y. Yavorsky, M. Gradhand, M. Czerner, M. Winkler, J. König, H. Böttner, I. Mertig and P. Zahn, *Phys. Rev. B* **86**, 085323 (2012).

**Mater. Res. Soc. Symp. Proc. Vol. 1543 © 2013 Materials Research Society**
**DOI: 10.1557/opl.2013.940**

# Giant Thermoelectric Effect in Graded Micro-Nanoporous Materials

Dimitrios G. Niarchos[1,*], Roland H. Tarkhanyan[1] and Alexandra Ioannidou[1]

[1]Institute for Advanced Materials, Physicochemical Processes, Nanotechnology & Microsystems, Department of Materials Science, NCSR "Demokritos", Athens, Greece

## ABSTRACT

In this work we report on opportunities for a colossal reduction in lattice thermal conductivity (LTC) of graded micro-nanoporous structures with inhomogeneous porosity which leads to the considerable improvement in thermoelectric figure of merit ZT. We employ the effective medium theory to calculate the LTC of a porous media with hole pores of variable radius and show that porous materials with inhomogeneous porosity are expected to have stronger reduction (about 30 times!) in thermal conductivity than those with pores of equal sizes. Such a reduction is caused by enhanced scattering of thermal phonons with the pore boundaries. We have studied the variations of the LTC as a function of porosity, pore sizes, geometry and the number of pore groups with different sizes. Our theoretical results show excellent agreement with experimental data.

## INTRODUCTION

In the last two decades, a great effort has been made to enhance the range of high-performance thermoelectric (TE) materials for industrial applications [1,2]. The key ideas for improving the efficiency of TE devices are connected with the enhancement of the power factor $P = \sigma S^2$ ($\sigma$ - electrical conductivity, $S$ - Seebeck coefficient) and reduction in the thermal conductivity $K=K_e+K_L$, where $K_e$ is the contribution of the free charge carriers and $K_L$ is the lattice thermal conductivity (LTC). Recent theoretical and experimental results [3-7] show that the LTCs of thin nanocomposite films can be reduced by orders of magnitude with respect to bulk material values. At present, semiconductor based quantum well superlattice structures and nanowires exhibit highest power factor and lowest LTC which leads to the significantly higher thermoelectric figure of merit $ZT = PT / K$ compared to that of the bulk materials [8-10]. However, in the technology of thermoelectric materials for general applications, the devices based on the bulk materials are more preferable.

In recent work of M. Dresselhaus et al [11] was theoretically proposed and experimentally proved the enhancement of the power factor (but not the thermal conductivity) in modulation doped silicon germanium alloy nanocomposites leading to ZT of 1.3 at 900°C. They have used $Si_{95}Ge$ as the matrix and $Si_{70}Ge_{30}P_3$ as the nanaparticles. Modulation-doping approach in thermoelectrics has been introduced in a previous work of the same group of authors [12]. This approach is based on well known in semiconductor physics phenomenon of spatial separation of

---

*Presenting Author. E-mail: dniarchos@ims.demokritos.gr

the charge carriers from their ionized parent atoms which leads to the reduction of the charge scattering and consequently to higher electrical conductivity σ. Using this approach, σ has been improved by 50%. On the other hand, the enhancement in σ leads to the increase in the electronic part of the thermal conductivity $k_e$ which negatively affect the figure of merit ZT. The enhancement in $k_e$ has been compensated by reduction in lattice thermal conductivity of the doped nanoparticles, since they are less thermal conductive than the matrix. The highest σ/K value for the modulation-doped samples happens at 35% of nanoparticles, which is 54% higher than σ/K value of equivalent uniform doped nanocomposite.

In the experimental work [13] it has been shown that in scutterudite compound $CoSb_3$ composed of large and small crystal grains there is a possibility for simultaneously enhancement of the power factor and a reduction of the thermal conductivity if the mass ratio of small particles is 30%.

Very recently the group of Prof. M. Kanatzidis achieved a record high ZT value of ~2.2 at 915K by introduction of 1-17nm SrTe nanocrystals and 0.1-1μm micrograins in Na-doped PbTe matrix[14]. They have used a "panoscopic" hierarchical architecture approach for integrated phonon scattering across multiple length scales from atomic-scale lattice disorder and nanoscale precipitates in the matrix to mesoscale grain boundaries. It is evident that in this way, more extensive phonon scattering and, consequently, stronger reduction in lattice thermal conductivity can be achieved than in the case of nanostructuring alone. However, in this work, like in most bulk materials, the optical phonons are ignored for the lattice thermal conductivity. On the other hand, applying first-principles calculations to lead telluride, Gang Chen et al [15] have shown that the optical phonons are important because they comprise over 20% of the lattice thermal conductivity and also they provide strong scattering channels for acoustic phonons, which is crucial for the low thermal conductivity.

Among various types of thermoelectric materials, porous media play an important role and represent a highly dynamic research area that aims at the creation of the next generation of efficient solid-state thermoelectric energy conversion devices[16]. Such materials have great potential for applications; the challenge is to lower LTC without reduction of the power factor. Since the pioneering work of Maxwell-Garnett [17], there have been many studies on thermoelectric properties of porous materials (see, for example, [18-25] and citations therein). It has been shown that the main mechanism for the reduction in LTC of composite materials is the scattering of phonons with pore boundaries and film interfaces. The presence of pores, indeed, decreases also the electrical conductivity compared to that in the bulk material at zero porosity. On the other hand, the band-bending at pore-medium interfaces produces potential barriers, and scattering at these barriers blocks low-energy charge carriers. The presence of traps as well as the energy-filtering effect at pore boundaries leads to the enhancement in the thermopower. This effect, together with the reduction in thermal conductivity, results in the improvement of the figure of merit over the one in the bulk material.

The purpose of this work is an investigation of opportunities for the reduction in thermal conductivity in porous structures with inhomogeneous porosity, as well as a discussion of several interesting peculiarities of thermoelelectric effects in such composites resulting in the improvement of the figure-of-merit ZT over the one in the bulk material.

The work consists of theoretical and experimental parts.

## THEORY

We model the porous material as a composite consisting of $N$ alternating layers of various thicknesses $l_n$, $n=1,2,...N$. In each layer there is a periodic three- dimensional cubic lattice of spherical hole pores with one hole per a cell, fixed lattice period $a_n$ and pore diameter $d_n$. The latter is assumed to be different in distinct layers while lattice period may be the same. Such a material behaves like a multilayer system consisting of layers of identical host materials with graded inhomogeneous porosity (Fig.1). Note that the lattice period can be expressed as $a_n = d_n + \delta_n$, where $\delta_n$ is the distance between adjacent pores (neck). We assume that $d_n < a_n < l_n$. One more important length scale is the averaged mean free path (MFP) of thermal phonons $\Lambda_b$ in the bulk around the pores. In crystalline materials, $\Lambda_b$ depends on many factors such as the temperature, thickness of the sample, phonon frequency, doping level etc. In the context of this paper, we do not aim at analyzing all these dependences in detail; rather, we refer the reader to the original experimental and theoretical publications [6, 26-29], the main subject of which is the reduction in thermal conductivity of thin films due to the shortening of the bulk phonon MFP as a result of scattering at the film boundaries. Note that the main difference of the current model compared to previous ones (e.g., [21]), is the presence of "graded" inhomogeneous porosity in the composite. Inhomogeneous porosity leads, as we will see, to the considerable reduction in the thermal conductivity of the structure.

Let us choose a Cartesian coordinate system with z-axis perpendicular to the plane of the layers, so as the first layer with $n=1$ occupies the space region $0 \le z \le l_1$, $0 \le x, y \le L_{x,y}$, where $L_{x,y}$ are the lengths of the sample in the $x, y$ directions, and the n-th layer occupies the region $l_{n-1} \le z \le l_n$.

By definition, the volume fraction (porosity) of the $n^{\text{th}}$ layer is

$$\phi_n = \frac{4}{3}\pi\left(\frac{d_n}{2}\right)^3 c_n = \frac{\pi}{6} c_n d_n^3, \tag{1}$$

where $c_n \equiv a_n^{-3}$ is the number density (concentration) of the pores. Since $d_n/a_n<1$, it is evident that the porosity is limited from above: $\phi_n<\pi/6$.

Assuming that the thermal conductivity of the pores is negligible (this assumption is only true if the heat radiation across pores and convection through pores can be neglected), the LTC for a single layer in the frequency independent (gray medium) approximation can be presented in the form[30]

$$\frac{K_n}{K_b} = \frac{1-\phi_n}{1+\Lambda_b / d_n \beta(\phi_n)}, \tag{2}$$

where

$$\beta(\phi_n) = (\pi / 6\phi_n)^{1/3} - 1, \tag{3a}$$

if the MFP $\Lambda_{pn}$ corresponding to the scattering at the pore/medium interfaces is equal to the distance between adjacent pores $\delta_n$. In the case when $\Lambda_{pn}$ does not equal to $\delta_n$, the expression for the function $\beta(\Box_n)$ is modified to the form

$$\beta(\phi_n) = 2/3\phi_n. \tag{3b}$$

Since we deal with large pore spacing compared to dominant phonon wavelength ($\delta_n/\lambda >> 1$), the effects of multiple scattering and the interference between phonon waves scattered by neighboring pores can be neglected; besides, the phonon-pore collisions can be considered as

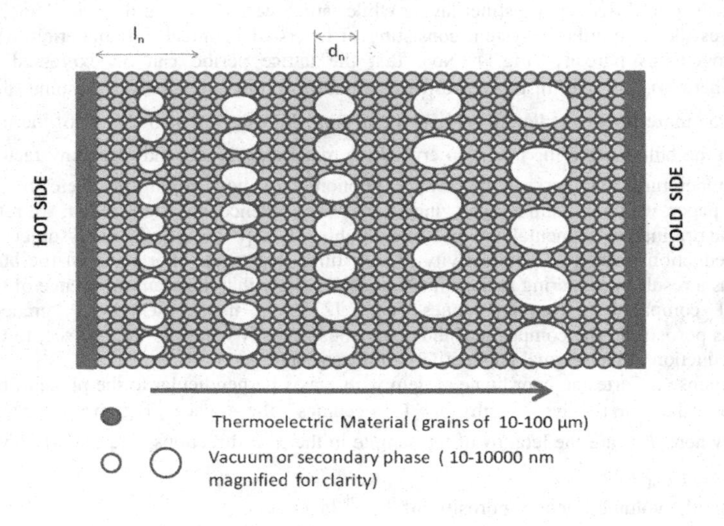


**Figure 1**. Schematic illustration of the multilayer configuration with layers of different porosity (graded porous material). Each layer contains a concentration of periodically distributed pores of the same size (only one row of such particles is shown).

independent scattering events.

As to the effective LTC of the composite with inhomogeneous porosity, it is given by [30]

$$\frac{1}{K_L} = \frac{1}{L_z} \sum_{n=1}^{N} \frac{l_n}{K_n}. \tag{4}$$

Let us consider now a particular case of the composite in which 1) a layer thickness $l_n = const \equiv l$, 2) both the concentrations and pore sizes in different layers are changing in such a way that the porosity of all the layers remains the same: $\phi_n = const \equiv \varphi$.

Furthermore, this case will be refereed as *special case of homogeneous porosity*. In this case from Eqs.(2) and (4) we obtain

$$\frac{K_L}{K_b}=\frac{1-\varphi}{1+\Lambda_b\beta^{-1}(\varphi)\langle d^{-1}\rangle}, \tag{5a}$$

where

$$\langle d^{-1}\rangle=\frac{1}{N}\sum_{n=1}^{N}\frac{1}{d_n}. \tag{5b}$$

Assuming that the distribution of the pore diameters is given, e.g., according to the relation

$$d_n=d_{\max}/n!, \tag{6}$$

where $d_{\max}$ is the diameter of the biggest pores (i.e., the pores in the first layer), one can obtain

$$\langle d^{-1}\rangle=d_{\max}^{-1}F(N),\quad F(N)=\frac{1}{N}\sum_{n=1}^{N}n!, \tag{7}$$

where $F(N)$ is a very quickly increasing discrete function of N (see Table I). Obviously, this fact results in effective reduction of the LTC with increasing of the number of pore groups with different sizes.

**Table I.** The values of the discrete function $F(N)$ and of the ratio $K_L/K_b$ for $N$=1-5.

| | $K_L/K_b$ | | | | |
|---|---|---|---|---|---|
| N | **1** | **2** | **3** | **4** | **5** |
| F(N) | **1** | **1.5** | **3** | **8.25** | **30.6** |
| φ=35% | 0.051 | 0.035 | 0.018 | 0.0067 | 0.0018 |
| φ=20% | 0.148 | 0.105 | 0.056 | 0.021 | 0.0058 |
| φ=10% | 0.276 | 0.204 | 0.115 | 0.048 | 0.013 |

By setting, e.g., $\Lambda_b$=300nm, $d_{max}$=180nm, and using Eqs (5a)-(7), at $\varphi$=35% we obtain $K_L/K_b=0.051$ and $1.8\times10^{-3}$ for N=1 and 5, respectively (see Table I), Thus, the presence of the pores with five different diameters leads to the reduction in the magnitude of LTC in about 30 times! Except that, from Table I we see that an increasing $N$ from 3 to 5 at the same porosity leads to the reduction of the magnitude of LTC by one order while that from 1 to 3 -only about 3 times.

Note that in the case when the distance between adjacent pores (or the neck, $\delta_n$) is known from the experiment (see, for example, Ref. [31]), Eqs. (2) and (3a) can be replaced by

$$\frac{K_n}{K_b}=\frac{1-\phi_n}{1+\Lambda_b/\delta_n}. \tag{8}$$

Then for N-layer composite in the *special case of homogeneous porosiity* we obtain

$$\frac{K_N}{K_b} = \frac{1-\phi}{1+(\Lambda_b / N)\sum_{n=1}^{N} \delta_n^{-1}}. \tag{9}$$

In a particular case when the necks are changing in according to the law

$$\delta_n = \delta_{\max} / n!, \tag{10}$$

Eq. (9) gives

$$\frac{K_N}{K_b} = \frac{1-\phi}{1+F(N)\Lambda_b / \delta_{\max}}. \tag{11}$$

For definiteness and simplicity, furthermore we will consider n-type non-polar isotropic thermoelectric material with a single parabolic conduction band and will neglect possible changes in the carrier effective mass due to the band-bend at pore-medium interfaces. Usually, the porosity leads to a small reduction in electrical conductivity because of additional scattering of electrons at pore boundaries. On the other hand, using the fact that the thermopower is equivalent to entropy per carrier per charge, it is not difficult to show that the presence of porosity leads to a small enhancement in the absolute value of the Seebeck coefficient. Clearly, always there is a possibility for a situation when the power factor of the porous composite differs from that at zero porosity, for example, in only 10%, i.e.,

$$\sigma S^2 = 0.9\sigma_b S_b^2. \tag{12}$$

Furthermore for estimations of the figure of merit we will restrict ourselves to the consideration of the case when condition (12) is assumed to be fulfilled.

## COMPARISON WITH EXPERIMENTAL RESULTS

As a proof of concept of our theory we present two examples from literature, where our approach gives better results that other models, thus confirming the basis of our approach.

Example 1. Phononic nanomesh Structures [7]

Phononic nanomesh structures of Si membranes using nanolithographic techniques are very good candidates for efficient thermoelectrics mainly due to the large thermal conductivity reduction. In Fig. 2 the studied structures are shown; the corresponding parameters and the experimental and simulation data are given in Table II.

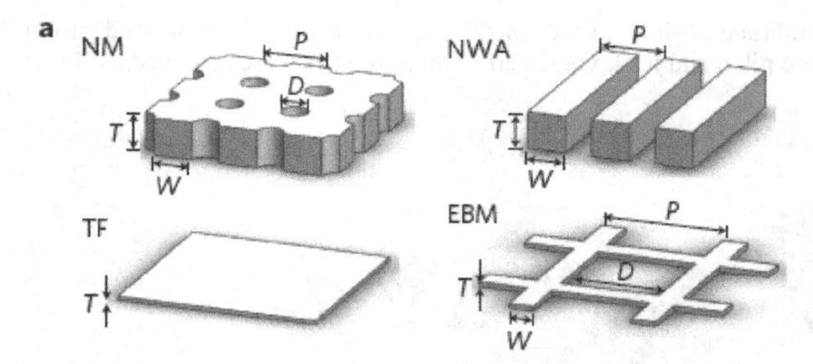


**Figure 2.** Geometry of the two nanomesh films (NM1 and NM2) and the two reference systems (TF, EBM) ( after [7]).

**Table II.** Dimensions of the nanomesh phononic structures together with experimental and theoretically calculated lattice thermal conductivity reduction.

| T (nm) (Thickness) | D (nm) Diameter | W(nm) Neck | $K/K_B$ (Theory) | | EUKEN | $K/K_B$(exp) | φ | Comment |
|---|---|---|---|---|---|---|---|---|
| | | | $\Lambda_B$=170 nm | $\Lambda_B$=300 nm | | $K_B$= 17 $W/m^2K$ | % | |
| 25 | | | 1 | 1 | 1 | 1 | 0 | |
| 22 | 270 | 115 | 1 | 1 | 1 | 1 | ~0 | We have to use $\Lambda_B$ = 170 nm for a thin ribbon of 50 nm |
| 22 | 16 | 23 | 0,105 | | 0,85 | 0,11 | 10.5 | |
| 22 | 11 | 18 | 0,107 | | 0,70 | 0,11 | 22 | |

From the table above it is apparent that modeling with our equations we can reproduce the experimental data although the Euken model fails completely.

Example 2. Holey Si [31]

Results have been obtained in a new type of nanostructure, holey silicon (HS), where high density nanoscopic holes are created in thin, single-crystalline silicon membranes. These HS nanostructures exhibit good mechanical strength and reproducibly low thermal conductivity while maintaining sufficient electrical quality. HS was prepared using either nanosphere lithography (NSL) or block copolymer (BCP) lithography, yielding holes with pitches of 350, 140, and 55 nm corresponding to diameters of holes of 198, 81 and 32 nm, neck values of 152, 59 and 23 and constant porosity of φ = 35 %.

In Fig.3 a holey Si membrane is shown used for the experiments for thermal conductivity measurements. By varying the pitch and neck values different porosities were achieved as shown in table III.

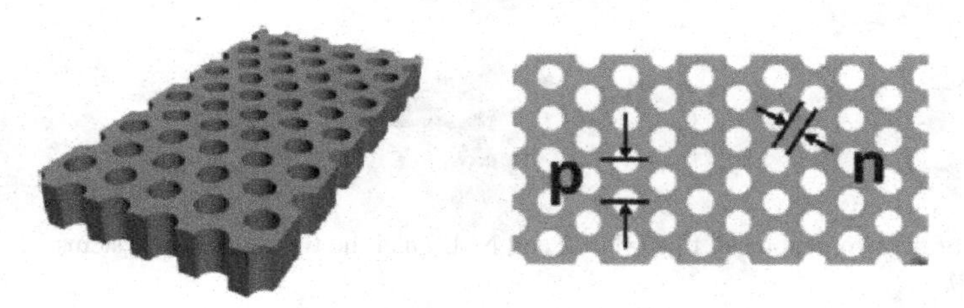


**Figure 3.** Holey Si nanostructures: p is the pitch and n is the neck used in the calculation below.

We have used our equations for discrete neck with $\varphi$=35%, bulk value of thermal conductivity of the membrane of thickness ~100 nm is $K_b$ ~ 50.9 $W/m^2K$ . We find that for such a thickness the MFP of phonons is $\Lambda_b$~300 nm rather than a smaller value, e.g. 45 nm. In Table III results of experimental data are compared with the predictions of our model, showing excellent agreement and once again the failure of the Euken model, indicative that scattering of phonons with different wavelength is important for the calculation of the lattice thermal conductivity in the nanomesh structures.

**Table III**. Diameters and neck values for Si membrane for $\varphi$=35 %. Two values for $\Lambda_b$ are used in order to show its importance for the reduction of the thermal conductivity.

| Diameter | Neck | $K/K_B$ (Theory) | | EUKEN | $K/K_B$(exp) | Comment |
|---|---|---|---|---|---|---|
| | | $\Lambda_B$=300 nm | $\Lambda_B$=45 nm | | $K_B$ =50.9 $W/m^2.K$ | |
| 198 | 152 | 0,218 | 0,501 | 0,55 | 0,205 | There is no way to reproduce the experimental data with $\Lambda_B$=45 nm (PRL-2013) |
| 81 | 59 | 0,107 | 0,369 | 0,55 | 0,139 | |
| 32 | 23 | 0,046 | 0,219 | 0,55 | 0,041 | |
| 3-layer | | 0,084 | 0,324 | N/A | N/A | Clear advantage of layering |

Unfortunately no experimental data exist with graded porosity to test the validity of our approach. The recent "panoscopic" approach by the Kanatzidis group [14], confirms our model and our first attempt to simulate the ZT enhancement predicts a 20 % increase in ZT even though we used crude data for the matrix and the inclusions. Preliminary data by the authors[32] in the system $CoSb_3$+ $SiO_2$ inclusions of various sizes ranging from 70, 150 and 400 nm and with a constant porosity of 25 % are consisted with our model where the holes are substituted with nano inclusions of $SiO_2$ with a thermal conductivity of 0,3, 0,9 and 1,3 $W/m^2K$ correspondingly, much lower than the matrix thermal conductivity.

From the data in the examples above we conclude that a) The reduction of the lattice thermal conductivity is significant upon nanostructuring and strongly dependent of the phonon mean free path; b) Based on our estimates but also from literature data the reduction of the electrical conductivity is much smaller and for Silicon is reduced by at most 10 % mainly due to the small value of the electron mean free path, which is of the order of 1-10 nm, and c) An increase of the ZT by at least an order of magnitude can be achieved.

## CONCLUSIONS

By modeling a porous composite material with inhomogeneous porosity as a combination of homogeneously porous layers of different porosity, we have derived simple expressions for the thermal conductivity as functions of porosity, pore sizes and the number of the pore groups $N$ with different characteristic sizes. We have shown that the reduction in the lattice thermal conductivity of such a composite is much more effective than that in the case of homogeneous porosity. Even in special case, when both the concentrations and pore sizes in different layers are changing in such a way that the porosity of all the layers remains the same, the size effect achieved in the structure leads to the colossal reduction in the LTC (about 30 times at N=5) as compared to the materials with pores of identical size. A comparison of our theoretical results with existing experimental results shows excellent agreement.

Obviously, the porosity always leads to the reduction in electrical conductivity and simultaneously to the enhancement in the absolute value of the Seebeck coefficient, so as the change in the value of the power factor due to porosity can be neglected. Thus, the main physical mechanism for an enhancement in thermoelectric figure-of-merit of a porous material remains the reduction of the thermal conductivity due to effective phonon scattering on the pore boundaries at the presence of the size effect.

Higher ZT values can be obtained if more than one single layer with either different hole diameters and neck values with the same porosity, or with same hole diameters and neck values but with different porosity will be employed thus achieving, depending on the material, "giant ZT values" of very good bulk thermoelectric materials. Further optimization is underway to optimize the parameters for all the known classes of thermoelectric materials, thus opening the door for more applications.

## ACKNOWLEDGMENTS

This work is supported by project NEXTEC of EU.

## REFERENCES

1. M.S. Dresselhaus, G. Chen, M.Y. Tang, R.G. Yang, H. Lee, D.Z. Wang, Z.F. Ren, J. P. Fleurial, P. Gogna, *Advanced Materials* **19,** 1043-1052 (2007).
2. J.F. Li, W.S. Liu, L.D. Zhao, M. Zhou, *.Nature Asia Mater.* **2,** 152–158 (2010).
3. L. D. Hicks, M. S. Dresselhaus, *Phys. Rev. B* **47,** 12727 (1993).

4. Y. S. Ju, K. E. Goodson, *Appl. Phys. Lett.* **74,** 3005- 3007 (1999).
5. G. Chen, M. S. Dresselhaus, G. Dresselhaus, J. P. Fleurial, T. Caillat, *Int. Mater. Rev.* **48,** 45- 66 (2003).
6. W. Liu, M. Asheghi, *Appl. Phys. Lett.* **84,** 3819-3821 (2004) .
7. J-K. Yu, S. Mitrovic, D. Tham, J. Varghese, J. R. Heath, *Nature Nanotech.* **5,** 718–721 (2010).
8. R. Venkatasubramanian, E. Siivola, T. Colpitts, B. O'Quinn, *Nature* **413,** 597-602 (2001).
9. T. C. Harman, P. J. Taylor, M. P. Walsh, B. E. LaForge, *Science* **297,** 2229-2239 (2002).
10. M. Maldovan, *J. Appl. Phys.* **110,** 034308 (2011).
11. Bo Yu, M. Zebarjadi, Hui Wang, K. Lukas, Hengzhi Wang, D. Wang, C. Opeil, M. Dresselhaus, G. Chen, Z. Ren, *Nano Lett.* **12**, 2077 (2012).
12. M. Zebarjadi, G. Joshi, G. Zhu, B. Yu, A. Minnich, Y. Lan, X. Wang, M. Dresselhaus, Z. Ren, and G. Chen, *Nano Lett.* **11**, 2225 (2011).
13. S. Katsuyama, F. Maezawa, T. Tanaka, *J. Physics: Conference Series* **379,** 012004 (2012).
14. K. Biswas, J. He, I. Blum, C.Wu, I. Hogan, D. Seidman, V. Dravid and M. Kanatzidis, *Nature* **489**, 414 (2012).
15. Z.Tian, J. Garg, K. Esfarjani,T. Shiga, J. Shiomi, G. Chen, *Phys.Rev.B* **85**, 184303 (2012).
16. H. Lee, D. Vashaee, D. Z. Wang, M. S. Dresselhaus, Z. F. Ren, G. Chen, *J. Appl. Phys.* **107,** 094308 (2010).
17. J.C. Maxwell-Garnet, *Philos. Trans. R. Soc. London* **203,** 385-420 (1904).
18. H.W. Russel, *J. Amer. Ceram. Soc.* **18,** 1-5 (1935).
19. D. J. Bergman, O. Levy, *J. Appl. Phys.* **70,** 6821-6833 (1991).
20. R. Landauer, *Electrical Transport and Optical Properties of Inhomogeneous Media*, (AIP, New York, 1978) pp. 2–45.
21. I. Sumirat, Y. Ando and S. Shimamura, *J. Porous Mater.* **13,** 439–443 (2006).
22. D. Song, G. Chen, *Appl. Phys. Lett.* **84,** 687-689 (2004).
23. R. G. Yang, G. Chen, M. S. Dresselhaus, *Nano Lett.* **5,** 1111-1115 (2005).
24. D. Vashaee, A.Shakouri, *Phys. Rev. Lett.* **92,** 106103 (2004).
25. S.V. Faleev, F. Leonard, *Phys. Rev.* **B77,** 214304 (2008).
26. J. Callaway, *Phys. Rev.* **113,** 1046-1051 (1959).
27. M. G. Holland, *Phys. Rev.* **132,** 2461-2471 (1963).
28. A. Majumdar, *J. Heat Transf.* **115,** 7-16 (1993).
29. G. Chen, *Phys. Rev.* **B57,** 14958 (1998).
30. R. H. Tarkhanyan and D.G. Niarchos, *Intern. J. of Thermal Sciences* **67**, 107-112 (2013).
31. J. Tang, H. Wang, D.H. Lee, M. Fardy, Z. Huo, T.P. Russell, P. Yang, *Nano Lett.* **10**, 4279 (2010).
32. A. Ioannidou, R. Tarkhanyan and D. Niarchos, To be published 2013.

Mater. Res. Soc. Symp. Proc. Vol. 1543 © 2013 Materials Research Society
DOI: 10.1557/opl.2013.920

# Computational modeling the electrocaloric effect for solid-state refrigeration

J.A. Barr[1], T. Nishimatsu[2], and S.P. Beckman[1]
[1]Department of Materials Science and Engineering, Iowa State University,
Ames, IA 50010, U.S.A.
[2]Institute for Materials Research (IMR), Tohoku University,
Sendai, 980-8577, Japan

## ABSTRACT

The electrocaloric effect holds promise for possible application in refrigeration technologies. There is much interest in this subject and experimental studies have shown the possibility for creating materials with a modest sized electrocaloric response. However, theoretical studies lag behind the experimental effort due to the lack of computational methods to accurately study the finite temperature response. Here the freely distributed feram, an effective Hamiltonian molecular dynamics method, is demonstrated for predicting the electrocaloric response of $BaTiO_3$.

## INTRODUCTION

A pyroelectric crystal experiences a spontaneous change in polarization as its temperature changes.[1] The electrocaloric effect (ECE) is the converse of this, i.e., the crystal experiences a spontaneous change in temperature when it's polarization changes. If the crystal is ferroelectric then the greatest pyroelectric and electrocaloric response occurs at the Curie temperature, where the crystal transforms from ferroelectric to paraelectric. It is possible to cycle the temperature and applied electric field to drive the conversion between thermal and electrical energies. One possible application is ECE based solid-state technologies for refrigeration.

Although there is great interest in seeing the development of ECE materials, the development of theoretical methods lags behind the experimental effort. First-principles molecular dynamics methods are not yet feasible, and thermodynamic modeling, such as the Ginzburg-Landau-Devonshire approach, requires substantial data for fitting. Here an effective Hamiltonian method will be demonstrated using both a direct and indirect molecular dynamics approach for the archetypical perovskite $BaTiO_3$ (BTO).

## METHODS

Here we use a first-principles based effective Hamiltonian model given by

$$H^{eff}=\frac{M^*_{dipole}}{2}\sum_{R,\alpha}\dot{u}^2_\alpha(R)+\frac{M^*_{acoustic}}{2}\sum_{R,\alpha}\dot{w}^2_\alpha(R)+V^{self}(\{u\})+V^{dpl}(\{u\})$$
$$+V^{short}(\{u\})+V^{elas,\mathrm{homo}}(\eta_1,...,\eta_6)+V^{elas,inho}(\{w\})$$
$$+V^{coup,\mathrm{homo}}(\{u\},\eta_1,...,\eta_6)+V^{coup,inho}(\{u\},\{w\})-Z^*\sum_R \varepsilon\bullet u(R).(1)$$

This model follows that developed in Refs. 2,3. The collective atomic motion is coarse-grained by local soft-mode vectors **u**(**R**) of each unit cell at **R** in a simulation supercell. For further efficiency, local acoustic displacement vectors **w**(**R**) are not treated in the molecular dynamics, but are optimized according to **u**(**R**). The details of this model and the methods are given in Refs. 4,5. The molecular dynamics method is encoded in the feram software package, which is distributed free of charge under the GNU public license at Ref. 6.

For the indirect approach the pyroelectric coefficient is predicted as a function of applied electric field and temperature. Then using the thermodynamical relationship

$$\Delta T = \frac{-1}{C_v}\int_{\varepsilon_1}^{\varepsilon_2} T\frac{\partial P_z}{\partial T}d\varepsilon_z \quad (2)$$

the adiabatic temperature change, ΔT, is determined.[7] In this expression $C_v$ is the heat capacity, $T$ is the temperature, $\frac{\partial P_z}{\partial T}$ is the pyroelectric coefficient, and the integral is taken over the applied electric field ranging from $\varepsilon_z = \varepsilon_1$ to $\varepsilon_z = \varepsilon_2$. The molecular dynamics simulation follows that outlined in Ref. 8. A periodic supercell contains N = 16 x 16 x 16 unit cells. The system is initialized with randomized soft-mode vectors. It is run through the temperature range of 250 K to 900 K using a temperature increment of 1 K/step. Each data point is thermalized for 20,000 time steps and is subsequently time averaged for 80,000 time steps using a time step of Δt=2 fs. A negative effective pressure is applied to the supercell to correct for the well known volume error endemic in density functional methods. A linear thermal expansion is also included such that the pressure is $p$ = -0.005T GPa. The data is collected at constant electric fields ranging from 25 to 310 kV/cm at an increment of 5 kV/cm.

For the direct approach, the supercell of size N = 96 x 96 x 96 is used. The supercell in the direct simulation must be significantly larger than the indirect to avoid the system-wide thermal fluctuations. A constant temperature canonical ensemble molecular dynamics calculation is run under an external electric field $\varepsilon_z$ and the system is allowed to equilibrate. The electric field is removed and at the same time the molecular dynamics simulation is changed to a constant energy micro-canonical ensemble, which is run using the leap-frog method. The thermal expansion is again simulated using a negative effective pressure $p$ = -0.005T GPa. With this method we do not have to use the experimental value of the heat capacity.

## RESULTS AND DISCUSSION

For the indirect approach the data is transformed to allow for the pyroelectric coefficient, $\frac{\partial P_z}{\partial T}$, to be expressed as a function of applied field at constant temperature. Approximating the heat capacity from the constant temperature experimental value[9], 2.53 J • cm$^{-3}$ • K$^{-1}$, it is possible to integrate Eqn. 2 for any range of applied fields. As discussed in Ref. 8, the applied fields will be held above the tricritical point, keeping the ferroelectric to paraelectric transformation second order to avoid the hysteresis effects and the latent heat of transformation. In the case of $BaTiO_3$, this lower bounds the applied field at around 60 kV/cm.

One advantage of the direct approach is that it can avoid the difficulty of calculating $\frac{\partial P_z}{\partial T}$, especially under weak applied external electric fields of $\varepsilon_z < 60$ kV/cm.

The trade off being that due to the coarse-graining and optimization treatment of **w(R)**, the heat capacity is underestimated, consequently, $\Delta T$ is overestimated by a factor of 5. To compare the direct and indirect molecular dynamics methods a field range of 100 kV/cm that is lower bounded at 60 kV/cm is selected. The adiabatic temperature change for the two is plotted in Fig. 1. Accounting for the degrees of freedom associated with the optical mode, it is found that the direct and indirect $\Delta T$ are comparable in 1/0.2 times.

The difference between the indirect method and the direct method with the 0.2 correction factor occurs for two reasons. First, for the direct method, the temperature dependent effective pressure is fixed by the canonical ensemble and does not vary when the electric field is removed. Second, in the temperature dependence of the heat capacity is automatically included for the direct calculation.

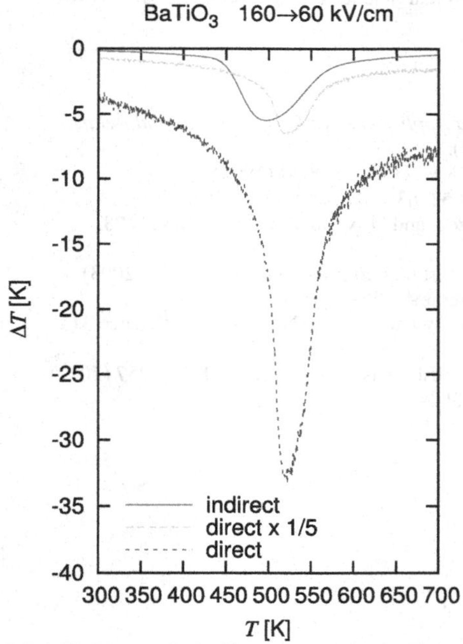


**FIG. 1. The electrocaloric effect compared for the direct measurement, the direct measurement with a correction factor, and the indirect measurement.**

## SUMMARY

We have demonstrated the application of feram, a freely distributed effective Hamiltonian molecular dynamics method[6] to predict the pyroelectric and electrocaloric response of the archetypical perovskite $BaTiO_3$. An indirect method is used that relies upon the thermodynamic expression in Eqn. 2 as well as a direct method where the system is switched from a canonical ensemble, where the system is equilibrated, to a micro-canonical ensemble, where the adiabatic temperature change can be directly measured when the field is switched. It is shown that these two methods yield equivalent results and therefore pose as comparable methods for predicting the ECE in perovskite ferroelectric ceramics.

## ACKNOWLEDGMENTS

This work was supported by JSPS KAKENHI Grant Number 23740230 and 25400314. Computational resources were provided by the Center for Computational Materials Science, Institute for Materials Research (CCMS-IMR), Tohoku University. We thank the staff at CCMS-IMR for their constant effort.

SPB and JAB were also supported by the US National Science Foundation under grant DMR-1105641.

## REFERENCES

1. M.E. Lines and A.M. Glass, *Principles and Applications of Ferroelectrics and Related Materials*, (Clarendon Press, Oxford, 1977).
2. R.D. King-Smith and D. Vanderbilt, Phys. Rev. B **49**, 5828-44 (1994).
3. W. Zhong and D. Vanderbilt, Phys. Rev. B **52**, 6301-12 (1995).
4. T. Nishimatsu, U.V. Waghmare, Y. Kawazoe, and D. Vanderbilt, Phys. Rev. B **78**, (2008).
5. U.V. Waghmare, E.J. Cockayne, and B.P. Burton, Ferroelectrics **291**, 187-96 (2003).
6. T. Nishimatsu, http://loto.sourceforge.net/feram/, (2007-2010).
7. A.S. Mischenko, Q. Zhang, J.F. Scott, R.W. Whatmore, and N.D. Mathur, Science **311**, 1270-1 (2006).
8. S.P. Beckman, L.F. Wan, J.A. Barr, and T. Nishimatsu, Mater. Lett. **89**, 254-257 (2012).
9. Y. He, Thermochimica Acta **419**, 135-41 (2004).

**Mater. Res. Soc. Symp. Proc. Vol. 1543 © 2013 Materials Research Society**
**DOI: 10.1557/opl.2013.678**

# Heat Transport between Heat Reservoirs Mediated by Quantum Systems

George Y. Panasyuk[1], George A. Levin[1], and Kirk L. Yerkes[1]
[1]Aerospace System Directorate, Air Force Research Laboratory,
Wright-Patterson Air Force Base, Ohio 45433, U.S.A.

## ABSTRACT

We explore a model of heat transport between two heat reservoirs mediated by a quantum particle. The reservoirs are modeled as ensembles of harmonic modes linearly coupled to the mediator. The steady state heat current, as well as the thermal conductance are obtained for arbitrary coupling strength and will be analyzed for the cases of weak and strong coupling regimes. It is shown that the violation of the virial theorem – the imbalance between the average potential and kinetic energy of the mediator – can be considered as a measure of the coupling strength that takes into account all the relevant factors. The dependence of the thermal conductance on the coupling strength is non-monotonic and displays a maximum. Temperature dependence of the heat conductance may reach a plateau at intermediate temperatures, similar to the classical plateau at high temperatures. We will discuss the origin of Fourier's law in a chain of macroscopically large, but finite subsystems coupled by the quantum mediators. We will also address the origin of the anomalously large heat current between the scanning tunneling microscope tip and the substrate in deep vacuum which was found in recent experiments.

## INTRODUCTION

A study of heat transfer through microscopic systems, such as molecules, nanotubes, and quantum dots, is one of central pursuits in modern physics contributing to both fundamental research and technological applications [1-2]. Our approach to study heat transport is based on the quantum Langevin equation [3-5] and the Drude-Ullersma model [4]. Dependencies of the derived expressions for the heat current $J_{th}$ and heat conductance $K$ on temperature $T$ and coupling strength $\gamma$ are analyzed. The derived quantities are used in discussion of Fourier's law origin and the interfacial heat transfer studied experimentally in [6]. This research is based on [7] with the following modifications: (i) major results (5)-(7) are presented here in a more general case when the couplings $\gamma_1$ and $\gamma_2$ can be different; (ii) figures 1 and 2 are drawn for different values of parameters than in [7] in order to show a wider scope of applicability of the presented theory; (iii) a possible deviation from Fourier's law is described (including Fig. 4); (iv) a better and, in our opinion, more convincing explanation of the results [6] is provided.

## THEORY

Our study is based on the total Hamiltonian $H_{tot} = H + H_{B1} + H_{B2} + V_1 + V_2$, where

$$H = \frac{p^2}{2m} + \frac{kx^2}{2},\quad H_{B\nu} = \sum_i \left[\frac{p_{\nu i}^2}{2m_{\nu i}} + \frac{m_{\nu i}\omega_{\nu i}^2 x_{\nu i}^2}{2}\right],\ \text{and}\ \ V_\nu = -x\sum_i C_{\nu i}x_{\nu i} + x^2\sum_i \frac{C_{\nu i}^2}{2m_{\nu i}\omega_{\nu i}^2} \tag{1}$$

are the Hamiltonians of the quantum system (the mediator), thermal reservoirs, and the interaction between the mediator and thermal reservoirs, respectively. Here $x$ and $p$ are the coordinate and momentum operators and $m$ and $k$ are the mass and spring constant of the mediator, $x_{\nu i}$ and $p_{\nu i}$ are the coordinates and momentum operators, whereas $m_{\nu i}$ and $\omega_{\nu i}$ are the masses and frequencies of the oscillators for the $i$th mode that belongs to the νth bath (ν = 1, 2). In addition, we employ the Drude-Ullersma model which assumes that in the absence of the interaction with the quantum system, each bath consists of uniformly spaced modes and introduces the following $\omega$ dependence for the coupling coefficients:

$$\omega_{\nu i} = i\Delta_\nu, \quad C_{\nu i} = \sqrt{\frac{2\gamma_\nu m_{\nu i}\omega_{\nu i}^2 \Delta_\nu D_\nu^2}{\pi(\omega_{\nu i}^2 + D_\nu^2)}}, \quad i = 1,2,...,N_\nu \tag{2}$$

Here $\Delta_\nu$ are the mode spacing constants, $D_\nu$ are the characteristic cutoff frequencies, and $\gamma_\nu$ are the coupling constants between a given reservoir and the mediator. We assume for simplicity that $D_1 = D_2 = D$. In the final results we make the limit $N_\nu \to \infty$ and $\Delta_\nu \to 0$. Formal solutions of the Heisenberg equations for $x_{\nu i}$ and $p_{\nu i}$ allow one to express these quantities through the unknown $x(t)$. Finally, using $x_{\nu i}$ in the dynamic equation for $x(t)$, one arrives at the quantum Langevin equation and its solution can be presented as

$$x(t) = \dot{g}(t)x(0) + \frac{1}{m}g(t)p(0) + \frac{1}{m}\int_0^t ds g(t-s)\eta(s), \tag{3}$$

where $g(t) = \sum_{n=1}^{3} g_n e^{-\mu_n t}$ with known $g_n$ and $\mu_n$ with $\mathrm{Re}(\mu_{1,2,3}) > 0$ (see [7] for details). Using the Heisenberg equations for $x_{\nu i}$ and $p_{\nu i}$, one finds that the rate of change the energy of the νth thermal reservoir is determined by the work our quantum system performs on this bath:

$$\frac{d}{dt}\sum_i \left\langle \frac{p_{\nu i}^2}{2m_{\nu i}} + \frac{m_{\nu i}\omega_{\nu i}^2 x_{\nu i}^2}{2} \right\rangle = -\langle P_\nu \rangle = \sum \frac{C_{\nu i}}{2m_{\nu i}} \langle p_{\nu i}x + xp_{\nu i} \rangle, \tag{4}$$

where the angular brackets means the ensemble averaging. In the steady-state the power acquired by one reservoir is taken from the other, so the heat current $J_{th}$ can presented as $J_{th} = \langle P_1 \rangle = -\langle P_2 \rangle = \langle P_1 - P_2 \rangle / 2$. Using the obtained solutions for $x_{\nu i}(t)$, $p_{\nu i}(t)$ and $x(t)$, one finds eventually the heat current $J_{th}$ and heat conductance $K = \lim_{\Delta T \to 0} J_{th} / \Delta T \quad (\Delta T = T_1 - T_2)$:

$$J_{th} = -\frac{2\hbar D^2 \gamma_1 \gamma_2}{\pi m \gamma} \sum_{n=1}^{3} g_n \mu_n^2 \int_0^\infty \frac{d\omega\omega[n_1(\omega) - n_2(\omega)]}{(D^2 + \omega^2)(\mu_n^2 + \omega^2)}, \; K = -\frac{\hbar^2 \gamma_1 \gamma_2 D^2}{2\pi k_B^2 T^2 m \gamma} \sum_{n=1}^{3} g_n \mu_n^2 \int_0^\infty \frac{d\omega\omega^2 \operatorname{cosech}^2(\hbar\omega\beta/2)}{(D^2 + \omega^2)(\mu_n^2 + \omega^2)}, \tag{5}$$

where $\gamma = \gamma_1 + \gamma_2$, $\beta = 1/k_B T$, $n_\nu(\omega) = 1/[\exp(\hbar\omega\beta_\nu) - 1]$, and $T_\nu$ is the temperature of the νth bath. For the high-temperature (classical) limit, when $\hbar |\mu_n| / k_B T_\nu << 1$, one has

$$J_{th} = -K(T_1 - T_2), \text{ where } K = \frac{k_B \gamma_1 \gamma_2}{\gamma m} \frac{2D^2}{2(D^2 + \omega_0^2) + \gamma D / m} \tag{6}$$

and for the low-temperature (quantum) limit, when $\hbar |\mu_n| / k_B T_\nu >> 1$, one finds

$$J_{th} = \frac{2\pi^3 k_B^4 \gamma_1 \gamma_2}{15\hbar^3 m^2 \omega_0^4}(T_1^4 - T_2^4) \quad \text{and} \quad K = \frac{8\pi^3 k_B^2 \gamma_1 \gamma_2 T^3}{15\hbar^3 \omega_0^4 m^2}. \tag{7}$$

**RESULTS AND DISCUSSION**

As we can show, the weak and strong coupling can be defined in terms of effective bath-particle interaction $\gamma_D = D^2\gamma/(D^2+\omega_0^2)$. If $\gamma_D/m << \omega_0$, the mediator can be described as an oscillator with small effective friction. The additional factor $D^2/(D^2+\omega_0^2)$ reflects the fact that if $D << \omega_0$, the baths' modes are effectively decoupled from the particle mode decreasing bath-particle interaction: it can be small even if $\gamma$ is not small. As is also shown, a small amplitude of the quantity, $\delta\widetilde{T}_{xp} = |T_p - T_x|/(T_p+T_x)$ where $k_B T_x = \langle kx^2\rangle$ and $k_B T_p = \langle p^2/m\rangle$ indicates a possibility to assign a certain temperature to the mediator. As we found, in addition to smallness of $(\gamma_D/m)/\omega_0$, temperatures of both reservoirs must be sufficiently high, i.e. $k_B T_{1,2} >> \hbar\gamma_D/m$. Only in this case the virial theorem ($\delta\widetilde{T}_{xp} << 1$) is satisfied and one can assign a certain temperature $T_M = T_x = T_p \approx [U(T_1)+U(T_2)]/2k_B$ for the mediator ($U(T) = (\hbar\omega_0/2)\coth(\beta\hbar\omega_0/2)$).

As our numerical simulations based on (5) reveal, $K$ may possess a maximum in its dependence on $\gamma$, as is shown in Figure 1. It appears when $\hbar\omega_0 \le k_B T$ and $\hat{\gamma}/D > 1$, $\hat{\gamma} \equiv \gamma/m$. Figure 2 shows $T$ dependence for the normalized heat conductance for different $D$ and for $\hbar\omega_0/k_B T = 10$. As we found, *K(T)* for $D/\omega_0 > 10$ is close to its $D \to \infty$ limit. The straight-line region corresponds to the low-*T* limit (7), which is the same for all curves. At large *T*, each curve reaches its classical plateau in accordance with (6). The physical origin of the intermediate plateau with a small value of $K \approx k_B\gamma^2 D^3/8\omega_0^4 m^2$ can be explained in the following way. When *T* increases above the Debye temperature $\Theta_D = D\hbar/k_B$, all the baths' modes are excited. At the same time, if *T* is still less than $\hbar\omega_0/k_B$ (which is always possible when $\omega_0 >> D$ as in Figure 2a), the mediator cannot be excited and, hence, cannot absorb energy from either bath and transfer it to the other one, leading to a small value of *K*. Moreover, because this situation stays unchanged when *T* changes from $\Theta_D$ to $\hbar\omega_0/k_B$, *K* does not depend on *T* noticeably and we have the plateau.

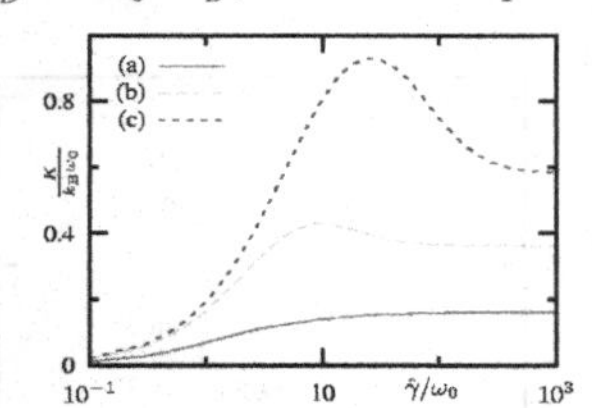


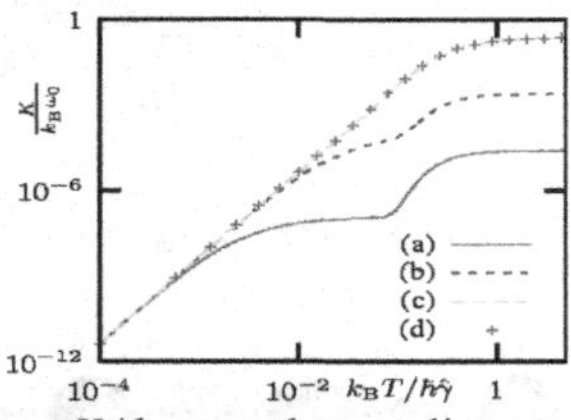


**Figure 1**. Dependence of the normalized heat conductance $K/k_B\omega_0$ on the coupling strength at $D/\hat{\gamma} = 3$. (a) $\hbar\omega_0/k_B T = 3$, (b) $\hbar\omega_0/k_B T = 1$, and (c) $\hbar\omega_0/k_B T = 0.3$. Here $\hat{\gamma} = \gamma/m$.

**Figure 2**. Temperature dependence for $K/k_B\omega_0$ at $\omega_0/\hat{\gamma} = 1$. (a) $D/\hat{\gamma} = 0.01$, (b) $D/\hat{\gamma} = 0.1$, (c) $D/\hat{\gamma} = 10$, and (d) $D/\hat{\gamma} = \infty$.

**Fourier's law**

As one can easily show, the difference in energies between two thermal reservoirs is

$\delta E \sim |(k_B T_1)^2/\hbar\Delta - (k_B T_2)^2/\hbar\Delta| \sim k_B^2 |T_1 - T_2|(T_1 + T_2)/2\hbar\Delta$. Taking into account that, as follows from (5) and (6), $J_{th} \sim k_B(\gamma/m)|T_1 - T_2|$, the characteristic time of mutual equilibration is $t_{eq} \sim \delta E/J_{th} \sim \Delta^{-1}$. With time, the average energies (or temperatures) of different independent baths' modes may deviate from each other in the absence of self-thermalization and a thermal reservoir cannot be characterized by a single temperature $T_\nu(t)$. However, if the observation time $t$ satisfies inequality $\tau_{micro} << t << t_{eq}$, where $\tau_{micro} \sim m/\gamma$ is a microscopic relaxation time, all modes will remain approximately in thermal equilibrium determined by the initial conditions. The same arguments allow us to extend the solution of the Hamiltonian (1) to the chain Hamiltonian whose graphical representation is shown in Figure 3. It is assumed that after preparation both TRs and all the subsystems in the state of thermal equilibrium, we turn on the couplings. On the time scale $t << t_{eq}$, the solution to the Hamiltonian (1) can be directly applied to the chain Hamiltonian in the form of energy conservation equations:

$$\partial_t E_n = J_{n,n-1} - J_{n,n+1} = K_{n-1,n}(T_{n-1} - T_n) - K_{n,n+1}(T_n - T_{n+1}). \quad (8)$$

After introducing a continuous coordinate $z = nd$, where $d$ is the distance between neighboring subsystems, (8) can be rewritten in the differential form of Fourier's law:

$$C(T)\partial_t T(z) = \partial_z[\kappa(z)\partial_z T(z)], \quad (9)$$

where $C = (dE/dT)/d$ is the specific heat of the chain, and $\kappa(T) = K(T)d$ is the thermal conductivity. As was already mentioned, one can expect that (8) or (9) are correct on the time scale $\tau_{micro} << t << t_{eq}$. It is worth mentioning that even in the case when thermalization in the subsystems does happen, Eq. (9) may not be correct if $K_n^- \equiv K_{n,n-1} \neq K_n^+ \equiv K_{n,n+1}$ (for example, if different mediators with different $\omega_0$'s are placed to the left and to the right of each subsystem). In this case, after introducing $z = nd$, $\delta K(z) = K(z + d/2) - K(z - d/2)$, and $\overline{K}(z) = [K(z + d/2) + K(z - d/2)]/2$, one arrives at the following equation instead of (9):

$$C(T)\partial_t T(z) = d^2\overline{K}(z)\partial_z^2(z) + d\delta K(x)\partial_z T(z). \quad (10)$$

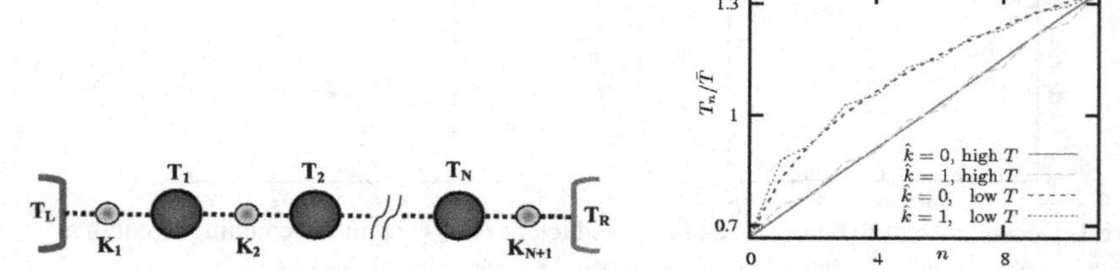


**Figure 3**. Chain of nanoparticles interconnected by quantum systems. Temperatures $T_{L,R}$ of the left and right reservoirs are fixed and temperatures $T_n$ of the nanoparticle can vary until the steady-state is established. Thermal conductances $K_n$ can vary from one connection to another.

**Figure 4**. Temperature profiles for high and low temperatures and different $\hat{k}$. Smooth curve correspond to cases when Fourier's law is applicable ($\hat{k} = 0$), whereas undulating curves describe temperature profiles for nonzero $\hat{k}$'s.

Figure 4 shows the steady-state temperature profile with N = 10 in situations when the heat conductance can be approximated as $K = \widetilde{K}T^r$. We assume that the chain is assembled in such a way that $\widetilde{K}_1 = \widetilde{K}_3 = ... = \widetilde{K}$, $\widetilde{K}_2 = \widetilde{K}_4 = ... = \widetilde{K}'$, and $\widetilde{K}' \neq \widetilde{K}$. In the high-temperature limit or in the case when $|T_R - T_L| << (T_R + T_L)/2 \equiv \overline{T}$, $r = 0$. In the low-temperature case, $r = 3$. Both cases are illustrated in Figure 4 for $|T_R - T_L|/\overline{T} = 2/3$ and $\hat{k} = 0$ or 1, where $\hat{k} \equiv (\widetilde{K}' - \widetilde{K})/K_{av}$ and $K_{av} = (\widetilde{K} + \widetilde{K}')/2$. Smooth curves correspond to $\hat{k} = 0$ when Fourier's law is applicable, and the undulating temperature profile is established if $\hat{k} = 1$.

**Interfacial heat transport**

In a recent experiment [6] the authors considered interfacial thermal transport between a gold substrate surface Au(1,1,1) and Pt/Ir tip with a CO molecule sitting on the tip with the carbon molecule attached to it. Using this molecule as an *in situ* thermometer, they found that despite the fact that the gap between the tip surface and substrate was only $\ell \approx 7$ Å, the temperature $T_M$ that corresponds to thermal vibrations of the molecule coincides with the substrate temperature $T_{sub}$. The authors of [6] tried to explain this observation invoking the model of the Planck emitter based on the acoustic mismatch model (AMM). In accordance to that model, the thermal current $J_{AMM}$ through the area $A$ is $J_{AMM} \approx A\pi^2 k_B^4 (T^4 - T_{sub}^4)/60\hbar^3 v_s^2$, where $v_s$ is the average speed of sound. It is assumed that emissivity is maximal (and equal to one) and $T_M = T$ where $T$ is the tip's end temperature. The ratio $J_{AMM}/J_{SB}$, where $J_{SB} = A\pi^2 k_B^4 (T^4 - T_{sub}^4)/60\hbar^3 c^2$ is the Stefan-Boltzmann law, is of the order of $10^{10}$ [8] and the authors of [6] tried to exploit this large value of $J_{AMM}/J_{SB}$ in order to explain why the molecule and substrate temperatures are close. The AMM, however, describes the heat transfer through the interface between two bulk material (and at least one of them is electrically insulating) and its relation to the experimental situation at hand is unclear from the physical point of view.

In our case, in order to estimate the ratio $R = J_{th}/J_{SB}$ for low temperatures, one can take into account experimental value $\omega_0 = 480\,\text{cm}^{-1}$, corresponding to the central frequency of Pt-C stretching vibration bond [9]. This energy feature was also observed in [6] and corresponds to oscillation motion of the whole CO molecule. It means that $m = m_C + m_O$ (masses of C and O atoms) in (7). As our numerical analysis shows, $R$ depends very weakly on $D$ and we replaced $D$ by the average value of the Debye frequencies for Pt and Au, which is $\overline{D} = 0.29\omega_0$. Thus, taking into account that $A = \pi d_C^2/4$, where $d_C = 1.4$ Å is a molecule size (in the direction perpendicular to the molecule's axis), one finds $R = 4 \cdot 10^9 (\hat{\gamma}/\omega_0)^2$. In order to estimate $\hat{\gamma}/\omega_0$, we observed that the FWHM of the Pt-C vibration mode (centered at $\omega_0 \approx 480\,\text{cm}^{-1}$) is approximately 60 $\text{cm}^{-1}$ [9]. It gives $R \approx 7 \cdot 10^7$ which corresponds to [10], where the heat current from evanescent waves was estimated by roughly eight orders of magnitude larger than $J_{SB}$. It is difficult to explain the above mentioned coincidence of the molecule and substrate temperatures by this relatively small value of $J_{th}$. As we found, however, the following, much simpler explanation for this fact is possible.

Indeed, disregarding the CO molecule at the tip end and taking into account that the tip is the Pt/Ir wire with the effective diameter at the tip's end $d_{tip} >> d_C, \ell$, one can estimate the near-field heat flux between the tip and substrate as [11] $S_{eva} = k_B^2(T^2 - T_{sub}^2)/6\hbar\ell^2$, where $\ell$ is the tip-substrate distance. Also, taking into account that the opposite end of the tip is at a temperature $T_0 > T_{sub}$, $S_{eva}$ is balanced by Fourier's flux $\kappa(T_0 - T)/L$, where $L$ is the tip's length and $\kappa$ is the thermal conductivity of the Pt/Ir tip. The resulting balance equation can be rewritten as $a(x^2 - 1) = (T_0/T_{sub}) - x$, where $a = k_B L T_{sub}/6\hbar\ell^2\kappa$ and $x = T/T_{sub}$. Taking into account that $\ell \approx 7$ Å, $L > 100\mu$m, $T_{sub} \geq 90$ K, and the value of $\kappa$ for the tip, one can find that $a >> 1$ and, correspondingly, $T \approx T_{sub}$ in accordance to the experiment [6]. Using (6), one can also estimate the ratio $r = J_{th}/J_{eva} = J_{th}/(S_{eva}\pi d_{tip}^2/4) = 0.35(\hat{\gamma}/\omega_0)(\ell/d_{tip})^2$. Thus, if $d_{tip} > 5$ nm and $\hat{\gamma} \leq \omega_0$, $r \leq 0.015$ and the additional heat current through the molecule can be, indeed, completely neglected: the only role of the molecule is to serve as an *in situ* thermometer.

## CONCLUSIONS

We derived expressions for the heat current and heat conductance between thermal reservoirs using the generalized quantum Langevin equation and the Drude-Ullersma model. The obtained expressions are analyzed for different temperatures, different coupling strength, and different Debye cutoff frequencies. The results are applied to a chain of macroscopically large but finite subsystems. As is shown, on a short time scale Fourier's law is validated. As is also shown, a deviation from Fourier's law due to different mediators placed on different sides of the subsystems is possible. An explanation of recent experimental results related to interfacial thermal transport is given.

## ACKNOWLEDGMENTS

The authors wish to acknowledge support from the Air Force Office of Scientific Research. One of the authors (G.Y.P.) is supported by the National Research Council Senior Associateship Award at the Air Force Research Laboratory.

## REFERENCES

1. Y. Dubi and M. Di Ventra, *Rev. Mod. Phys.* **83**, 131 (2011).
2. *Molecular Electronics*, edited by J. Jortner and M. Ratner (Blackwell Science, Oxford, 1997).
3. A.O. Caldeira and A.J. Leggett, *Physica* **121 A**, 587 (1983).
4. Th. M. Nieuwenhuizen and A. E. Allahverdyan, *Phys. Rev. E* **66**, 036102 (2002).
5. D. Segal, A. Nitzan, and P. Hanggi, *J. Chem. Phys.* **119**, 6840 (2003).
6. I. Altfeder, A.A. Voevodin, and A.K. Roy, *Phys. Rev. Lett.* **105**,166101 (2010).
7. G.Y. Panasyuk, G.A. Levin, and K.L. Yerkes, *Phys. Rev. E* **86**, 021116 (2012).
8. J.P. Wolfe, *Imaging Phonons: Acoustic Wave Propagation in Solids* (Cambridge University Press, New York, 1998).
9. G. Bylholder and R. Sheets, *J. Phys. Chem.* **74**, 4335 (1970).
10. A.I. Volokitin and B.N.J. Persson, *Rev. Mod. Phys.* **79**, 1291 (2007).
11. P. Ben-Abdallah and K. Joulain, *Phys. Rev. B* **82**, 121419(R) (2010).

Mater. Res. Soc. Symp. Proc. Vol. 1543 © 2013 Materials Research Society
DOI: 10.1557/opl.2013.955

# Detailed Theoretical Investigation and Comparison of the Thermal Conductivities of n- and p-type $Bi_2Te_3$ Based Alloys

Ö. Ceyda Yelgel, Gyaneshwar P. Srivastava
School of Physics, University of Exeter, Stocker Road, Exeter, EX4 4QL, United Kingdom

## ABSTRACT

In this work we present a detailed theoretical investigation of the thermal conductivities of n-type 0.1 wt.% CuBr doped 85% $Bi_2Te_3$ - 15% $Bi_2Se_3$ and p-type 3 wt% Te doped 20% $Bi_2Te_3$ - %80 $Sb_2Te_3$ single crystals. The thermal conductivity contributions arising from carriers, electron-hole pairs and phonons are computed rigorously in the temperature range 300 K $\leqslant T \leqslant$ 500 K. In agreement with available experimental measurements we theoretically find that the lowest total thermal conductivity is 3.15 W $K^{-1}$ $m^{-1}$ at 380 K for the n-type alloy and 1.145 W $K^{-1}$ $m^{-1}$ at 400 K for the p-type alloy. Stronger mass-defect scattering is found to be responsible for the lower thermal conductivity of the p-type alloy throughout the temperature range of the study.

## INTRODUCTION

Thermoelectric devices are capable of converting temperature differences into electric voltage and vice versa. The maximum efficiency of a thermoelectric material is determined by its dimensionless figure of merit ($ZT$):

$$ZT = \frac{S^2\sigma}{\kappa_c + \kappa_{bp} + \kappa_{ph}}T, \tag{1}$$

where $T$ is the absolute temperature, $S$ is the Seebeck coefficient, $\sigma$ is the electrical conductivity, and $\kappa_c$, $\kappa_{bp}$, $\kappa_{ph}$, are the carrier, electron-hole pair (bipolar) and phonon contributions of the thermal conductivity, respectively. One of the traditional ways to improve $ZT$ is making alloys of single crystals. By this method significant reduction can be gained for the phonon thermal conductivity and larger values of $ZT$ attained [1, 2]. Among various thermoelectric materials, $Bi_2Te_3$ based alloys are chracterised with reasonably large $ZT$ values near room temperature and further enhancement can be attempted by using diverse methods [3, 4, 5, 6].

In this present study we report a detailed theoretical investigation of the thermal conductivities of n-type 0.1 wt.% CuBr doped 85% $Bi_2Te_3$ - 15% $Bi_2Se_3$ and p-type 3 wt% Te doped 20% $Bi_2Te_3$ - %80 $Sb_2Te_3$ single crystals. Our theoretical thermal conductivity results are compared with the experimental values previously obtained by Hyun *et al.* [7] and Li *et al.* [8]. Various contributions of the conductivity are analysed to provide an explanation for the difference in the results for the n-type and p-type alloys. Moreover, the frequency dependence of phonon thermal conductivity is studied for the p-type alloy and compared to n-type alloy reported in our former work [9].

## THEORY

The total thermal conductivity in semiconductors is expressed as $\kappa_{\text{total}} = \kappa_{\text{c}} + \kappa_{\text{bp}} + \kappa_{\text{ph}}$ where the contributions are from carriers (electrons or holes, $\kappa_{\text{c}}$), electron-hole pairs (bipolar, $\kappa_{\text{bp}}$) and phonons ($\kappa_{\text{ph}}$).

*Carrier Thermal Conductivity*: The carrier thermal conductivity is determined by the Wiedemann-Franz law as [1, 2]

$$\kappa_{\text{c}} = \sigma \mathcal{L} T = \left(\frac{k_B}{e}\right)^2 \sigma T \mathcal{L}_0, \tag{2}$$

where $k_{\text{B}}$ is the Boltzmann constant, $\mathcal{L}$ is the Lorenz number and $\mathcal{L}_0$ is described in terms of the scattering parameter $r$ and the Fermi integral $F_i = \int_0^\infty \frac{x^i dx}{e^{(x-\zeta^*)}+1}$ as [10]

$$\mathcal{L}_0 = \frac{\left(r+\frac{7}{2}\right) F_{r+\frac{5}{2}}(\zeta^*)}{\left(r+\frac{3}{2}\right) F_{r+\frac{1}{2}}(\zeta^*)} - \left[\frac{\left(r+\frac{5}{2}\right) F_{r+\frac{3}{2}}(\zeta^*)}{\left(r+\frac{3}{2}\right) F_{r+\frac{1}{2}}(\zeta^*)}\right]^2. \tag{3}$$

*Bipolar Thermal Conductivity*: In narrow band-gap semiconductors the bipolar contribution to the total thermal conductivity becomes significant above room temperature [11, 12]. As reported in our previous works [9, 13] this contribution can be written as

$$\kappa_{\text{bp}} = F_{\text{bp}} T^p \exp(-E_{\text{g}}/2k_{\text{B}}T), \tag{4}$$

with $F_{\text{bp}}$ and $p$ regarded as adjustable parameters depending on doping type and $E_{\text{g}}$ is the energy band gap of a material.

*Phonon Thermal Conductivity*: By applying Debye's isotropic continuum model within the single-mode relaxation time scheme the phonon thermal conductivity is expressed as [14]

$$\kappa_{\text{ph}} = \frac{\hbar^2 q_D^5}{6\pi^2 k_B T^2} \sum_s c_s^4 \int_0^1 dx x^4 \tau \bar{n}(\bar{n}+1), \tag{5}$$

where $q_D$ is the Debye radius, $\hbar$ is the reduced Planck's constant, $c_s$ is the phonon speed for polarisation branch $s$, $\bar{n}$ is the Bose-Einstein distribution function and $x = q/q_D$. The phonon relaxation rate $\tau$ is contributed from several scattering mechanisms: boundary (bs), mass-defects (md), donor electrons (ep) or acceptor holes (hp), and anharmonic (anh). Expressions for $\tau_{\text{bs}}^{-1}$, $\tau_{\text{ep}}^{-1}$ and $\tau_{\text{hp}}^{-1}$ have already been given in detail in Refs. [9, 13, 14]. In semiconductor alloys the mass defect scattering of phonons arises from two different resources: isotopic point defects and mass difference due to alloying. Both types of mass defects can be expressed in the form [14, 15]

$$\tau_{qs}^{-1}(\text{md}) = \frac{\Gamma_{\text{md}}\Omega}{4\pi\bar{c}^3}\omega^4(qs), \tag{6}$$

where $\Omega$ is the volume of a unit cell, $\bar{c}$ is the average phonon speed and $\omega = cq$. Expressions for the isotopic and alloying contributions towards $\Gamma_{\text{md}}$, viz. $\Gamma_{\text{isotopes}}$ and $\Gamma_{\text{alloy}}$, are reported in our previous work already [9]. The anharmonic phonon scattering rate, following Srivastava's scheme [14], is expressed as

$$\begin{aligned}\tau_{qs}^{-1}(\text{anh}) &= \frac{\hbar q_D^5 \gamma^2}{4\pi\rho\bar{c}^2} \sum_{s's''\varepsilon} \left[ \int dx' x'^2 x''_+ [(1-\varepsilon+\varepsilon(Cx+Dx')] \frac{\bar{n}_{q's'}(\bar{n}''_+ + 1)}{(\bar{n}_{qs}+1)} \right. \\ & \left. + \frac{1}{2} \int dx' x'^2 x''_- [1-\varepsilon+\varepsilon(Cx-Dx')] \frac{\bar{n}_{q's'}\bar{n}''_-}{\bar{n}_{qs}} \right], \qquad (7)\end{aligned}$$

where $\gamma$ is the Grüneisen constant, $\rho$ is the mass density, $x' = q'/q_D$, $x''_{\pm} = Cx \pm Dx'$, $\bar{n}''_{\pm} = \bar{n}(x''_{\pm})$, $C = c_s/c_{s''}$, $D = c_{s'}/c_{s''}$, $\varepsilon = 1$ for momentum-conserving Normal processes, and $\varepsilon = -1$ for momentum-nonconserving Umklapp processes. The first and second terms in equation (7) are controlled by class 1 events $\boldsymbol{qs} + \boldsymbol{q}'s' \rightarrow \boldsymbol{q}''s''$ and class 2 events $\boldsymbol{qs} \rightarrow \boldsymbol{q}'s' + \boldsymbol{q}''s''$, respectively. The integration limits on the variables $x$ and $x'$ are derived from a detailed consideration of the energy and momentum conservation requirements and described in Ref. [14]. As we did in our previous works [9, 13], the phonon-phonon scattering is expressed in terms of the parameter defined as $F_{\text{3ph}} = (\frac{\gamma}{\bar{c}})^2$.

## RESULTS

Thermal conductivity calculations are performed for n-type 0.1 wt.% CuBr doped 85% $Bi_2Te_3$ - 15% $Bi_2Se_3$ and p-type 3 wt% Te doped 20% $Bi_2Te_3$ - %80 $Sb_2Te_3$ single crystals at temperatures from 300 K to 500 K and all the related parameters for the theoretical computation are compiled in Tab. 1.

By applying the Wiedemann-Franz law the carrier thermal conductivities of both n- and p-type doped $Bi_2Te_3$ based alloys are theoretically computed in the temperature range 300 K $\leqslant T \leqslant$ 500 K and presented in Fig. 1 (a). For comparision, experimental results studied by Hyun *et al.* [7] and Li *et al.* [8] are also presented in that figure. For the calculation of $\kappa_c$ the required electrical conductivity results are taken from our previously published works [9, 13]. In agreement with the experimental measurements, our theoretical results clearly show that $\kappa_c$ decreases consistently as the temperature increases for both n- and p-type doped alloys. The significantly lower carrier thermal conductivity values for the p-type alloy throughout the temperature range results directly as a consequence of its electrical resistivity being two times bigger than the n-type alloy. With the choice of the parameters presented in Tab. 1, we successfully reproduce the experimental measurements for the two different alloys and find the lowest values of 1.4 W $K^{-1}$ $m^{-1}$ and 0.6 W $K^{-1}$ $m^{-1}$, at 500 K, for the n- and p-type alloy, respectively.

The theoretically computed temperature dependence of the bipolar thermal conductivity is shown in Fig. 1 (b) for both the n- and p-type doped alloys. As expected from the Eq. (4), for both samples $\kappa_{\text{bp}}$ goes up exponentially as the temperature increases and becomes significantly important above room temperature. The smaller value of $\kappa_{\text{bp}}$ for the p-type alloy results from its narrower energy band gap at a given temperature, as discussed in Ref. [13]. The rise in $\kappa_{\text{bp}}$ from 300 K to 500 K is faster for the n-type alloy than for the p-type

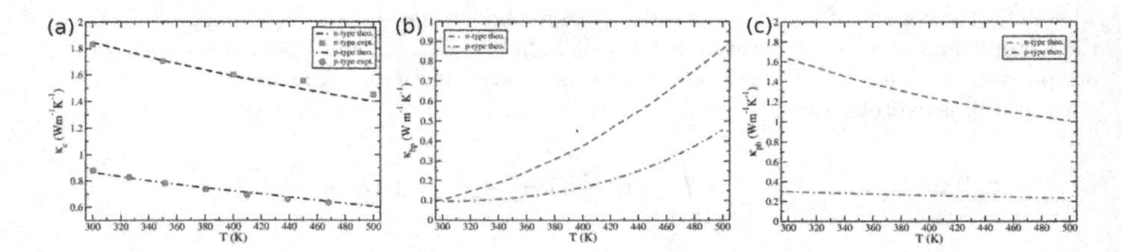

Figure 1: Temperature dependence of (a) the carrier thermal conductivity, (b) the bipolar thermal conductivity, (c) the phonon thermal conductivity for n-type 0.1 wt.% CuBr doped $(Bi_2Te_3)_{0.85}(Bi_2Se_3)_{0.15}$ and p-type 3 wt% Te doped $(Bi_2Te_3)_{0.20}(Sb_2Te_3)_{0.80}$ single crystals. The symbols represent the experimental results studied by Hyun *et al.* [7] and Li *et al.* [8].

alloy. At 500 K, the maximum value of the $\kappa_{\text{bp}}$ reaches to 0.86 W K$^{-1}$ m$^{-1}$ for the n-type alloy and by 0.456 W K$^{-1}$ m$^{-1}$ for the p-type alloy.

Figure 1 (c) shows the temperature dependence of the phonon thermal conductivity for n- and p-type alloys. For both type alloys we theoretically found that boundary and carrier-phonon scatterings are dominant only at low temperatures ($T < 100$ K), the phonon-phonon interaction becomes significant only at high temperatures ($T > 100$ K) and mass-defect scatterings play an important role throughout the temperature range. The p-type doped alloy has notably smaller $\kappa_{\text{ph}}$ value than the n-type alloy, with the $\kappa_{\text{ph}}(\text{n}-\text{type})/\kappa_{\text{ph}}(\text{p}-\text{type})$ ratio found to be 5.4 at 300 K and 4.34 at 500 K. From our theoretical calculations we find that compared to the n-type doped sample the p-type alloy is characterised by significantly larger mass-defect scatterings determined by $\Gamma_{\text{isotopes}}$ and $\Gamma_{\text{alloy}}$ parameters. Differences in these parameters upon alloy formation can easily be appreciated by noting that whereas Bi has no stable isotopes, Sb, Se and Te have two, six and eight isotopes, respectively. Also, the n-type alloying results in a much smaller mass difference ($M_{Te} - M_{Se} = 48.6$ amu) compared to the p-type alloying ($M_{Bi} - M_{Sb} = 87.2$ amu).

As demonstrated in Fig. 2 (a) the calculated total thermal conductivities of n- and p-type doped alloys successfully explain the experimental results obtained by Hyun *et al.* [7] and Li *et al.* [8]. The lowest value of $\kappa_{\text{total}}$ is found to be 3.15 W K$^{-1}$ m$^{-1}$ at 380 K for and 1.145 W K$^{-1}$ m$^{-1}$ at 400 K for the n- and p-type samples, respectively. From the theoretical calculations, we clearly establish that the smaller value of $\kappa_{\text{total}}$ for the p-type alloy throughout the temperature range results from its phonon conductivity being nearly five times smaller than that for the n-type alloy. This suggests that using p-type doped $Bi_2Te_3$ based alloy rather than n-type doped alloy is likely to give rise to a higher value of $ZT$.

The spectral analysis of the phonon thermal conductivity in the frequency space is represented in Fig. 2 (b) for the p-type doped $(Bi_2Te_3)_{0.20}(Sb_2Te_3)_{0.80}$ single crystal at several temperatures. Similar to the n-type doped $(Bi_2Te_3)_{0.85}(Bi_2Se_3)_{0.15}$ single crystal studied in Ref. [9] the spectrum becomes wider and the peak shifts to higher frequency when the temperature increases. This can be easily explained by noting that the energy of the dominant

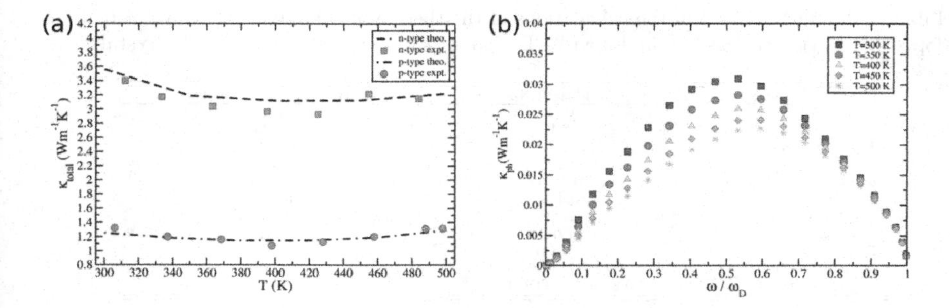


Figure 2: (a) Temperature dependence of the total thermal conductivity for n-type 0.1 wt.% CuBr doped $(Bi_2Te_3)_{0.85}(Bi_2Se_3)_{0.15}$ and p-type 3 wt.% Te doped $(Bi_2Te_3)_{0.20}(Sb_2Te_3)_{0.80}$ single crystals. The symbols represent the experimental results studied by Hyun *et al.* [7] and Li *et al.* [8]. (b) Frequency dependence of the phonon thermal conductivity for p-type 3 wt.% Te doped $(Bi_2Te_3)_{0.20}(Sb_2Te_3)_{0.80}$ single crystal at several temperatures where the Debye frequency is taken as $\omega_D$=17.62 THz.

phonon is directly proportional to crystal temperature: $\hbar\omega_{dom} \simeq 1.6k_BT$ [18]. Furthermore, it is theoretically found that the spectrum peaks at the frequency $\omega_D/1.92$ (where $\omega_D$ is the Debye frequency) and $\omega_D/1.66$ at 300 K and 500 K, respectively. Compared to the n-type doped alloy [9] the maximum value of $\kappa_{\text{ph}}$ occurs at a considerably lower frequency for the p-type doped alloy. This results from the smaller Debye frequency of 17.62 THz for the p-type alloy, compared to 18.92 THz for the n-type alloy sample [9].

## SUMMARY

The thermal conductivity contributions arising from carriers, electron-hole pairs, and phonons are theoretically studied for n-type 0.1 wt.% CuBr doped $(Bi_2Te_3)_{0.85}(Bi_2Se_3)_{0.15}$ and p-type 3 wt.% Te doped $(Bi_2Te_3)_{0.2}(Sb_2Te_3)_{0.8}$ single crystals and relevant parameters has been fitted to successfully reproduce the experimental measurements obtained by Hyun *et al.* [7] and Li *et al.* [8] in the temperature range 300 K $\leqslant T \leqslant$ 500 K. In this temperature range nearly three times smaller total thermal conductivity is attained for the p-type doped alloy due to its lower $\kappa_{\text{c}}$, $\kappa_{\text{bp}}$, and $\kappa_{\text{ph}}$ results. Comparatively significant reduction in $\kappa_{\text{ph}}$ for the p-type sample comes from much stronger isotopic and alloy scattering rates of phonons. Additionally, the phonon thermal conductivity - frequency spectrum is studied for the p-type doped sample and found that the spectrum becomes wider and the peak value shifts to higher frequencies as the temperature increases. Compared to n-type doped alloy, the highest value of $\kappa_{\text{ph}}$ takes place at smaller frequency owing to its lower Debye frequency.

Table 1: Parameters used in the calculations of the thermal conductivity of 0.1 wt.% CuBr doped $(Bi_2Te_3)_{0.85}(Bi_2Se_3)_{0.15}$ and 3 wt.% Te doped $(Bi_2Te_3)_{0.2}(Sb_2Te_3)_{0.8}$ single crystals.

| Property/Parameter | $(Bi_2Te_3)_{0.85}(Bi_2Se_3)_{0.15}$ | $(Bi_2Te_3)_{0.2}(Sb_2Te_3)_{0.8}$ |
|---|---|---|
| $E_g(0)$ (eV) | 0.21 | 0.09 |
| $r$ | 0.1 | 0.45 |
| $F_{bp}$ ($Wm^{-1}K^{-1}$) | $3.4\times10^{-4}$ | $13.0\times10^{-4}$ |
| $p$ | 1.0 | 1.0 |
| $q_D$ ($Å^{-1}$) | 0.7113 | 0.603 |
| $\Omega$ ($Å^3$) | 164.94 | 160.414 |
| $\bar{c}$ (m/s) [16] | 2611 | 2922 |
| $\Gamma_{isotopes}$ | 0.000112 | 0.025 |
| $\Gamma_{alloy}$ | 0.00447 | 0.05 |
| $\rho$ ($kg/m^3$) [17] | $7.7\times10^3$ | $6.69\times10^3$ |
| $F_{3ph}$ ($s^2/m^2$) | LT:$0.2\times10^{-5}$, HT:$0.4\times10^{-5}$ | $1.057\times10^{-5}$ |

## ACKNOWLEDGEMENT

Övgü Ceyda Yelgel is grateful for financial support from The Republic of Turkey Ministry of National Education through the Recep Tayyip Erdoğan University in Rize/Turkey.

## REFERENCES


[1] D. M. Rowe and C. M. Bhandari, 'Modern Thermoelectrics' (Reston Publishing Company, Virginia, 1983).
[2] D. M. Rowe, 'Thermoelectrics Handbook' (Taylor and Francis Group, London, 2006).
[3] B. Poudel, Q. Hao, Y. Ma, Y. Lan, A. Minnich, B. Yu, X. Yan, D. Wang, A. Muto, D. Vashaee, X. Chen, J. Liu, M. S. Dresselhaus, G. Chen, Z. Ren, Science **320** 634 (2008).
[4] R. Venkatasubramanian, E. Siivola, T. Colpitts, and O'Quinn, Nature **413** 597 (2001).
[5] V. Goyal, D. Teweldebrhan, and A. A. Balandin, App. Phys. Lett. **97** 133117 (2010).
[6] F. Zahid and R. Lake, App. Phys. Lett. **97** 212102 (2010).
[7] D. B. Hyun, J. S. Hwang, B. C. You, T. S. Oh, C. W. Hwang, J. Mat. Sci. **33** 5595 (1998).
[8] D. Li, R. R. Sun, X. Y. Qin, Intermet. **19**, 2002 (2011).
[9] Ö. C. Yelgel, G. P. Srivastava, Phys. Rev. B **85**, 125207 (2012).
[10] R. R. Heikes and R. W. Ure, Thermoelectricity, Science and Engineering (Interscience Publishers, New York, 1961).
[11] P. J. Price, Phil. Mag. **46**, 1252 (1955).
[12] C. J. Glassbrenner and G. A. Slack, Phys. Rev. **134**, A1058 (1964).
[13] Ö. C. Yelgel, G. P. Srivastava, J. Appl. Phys. **113**, 073709 (2013).
[14] G. P. Srivastava, 'The Physics of Phonons' (Taylor and Francis Group, New York, 1990).
[15] M. G. Holland, Phys. Rev. **134** A471 (1964).
[16] L. W. Silva and M. Kaviany, Int. J. Heat and Mass **47** 2417 (2004).
[17] D. R. Lide, 'CRC Handbook of Chemistry and Physics' (Taylor and Francis Group LLC, 87th Edition, 2007).
[18] J. M. Ziman, 'Electrons and Phonons' (Clarendon Press, Oxford, 1960).

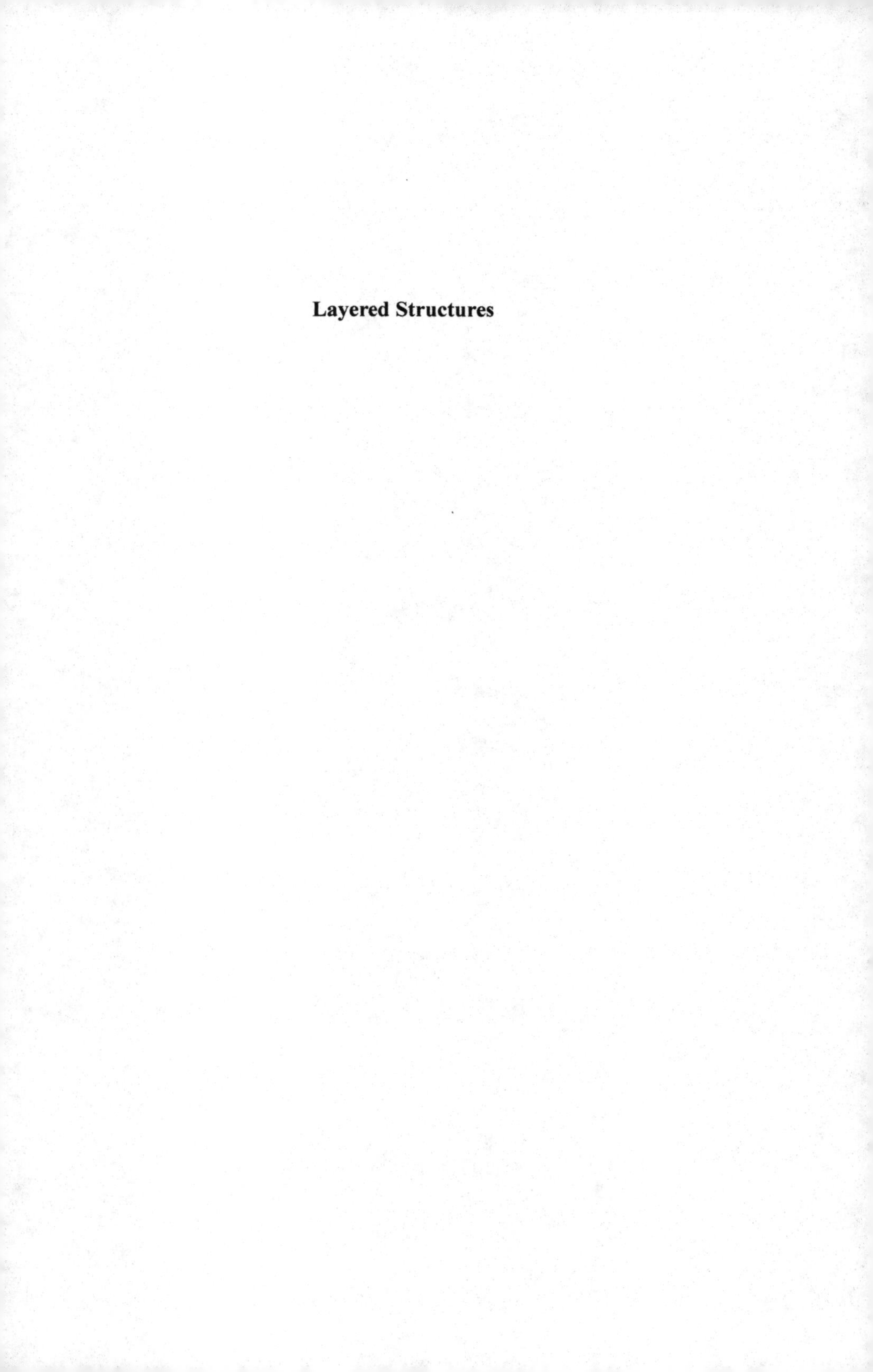

# Layered Structures

Mater. Res. Soc. Symp. Proc. Vol. 1543 © 2013 Materials Research Society
DOI: 10.1557/opl.2013.989

# Enhanced thermoelectric properties of Al-doped ZnO thin films

P. Mele[1], S. Saini[1], H. Abe[2], H. Honda[2], K. Matsumoto[3], K. Miyazaki[4], H. Hagino[4], A. Ichinose[5]

[1]Hiroshima University, Institute for Sustainable Sciences and Development, 739-8530 Higashi-Hiroshima, Japan,
[2]Hiroshima University, Graduate School for Advanced Sciences of Matter, 739-8530 Higashi-Hiroshima, Japan,
[3]Kyushu Institute of Technology, Department of Material Science, 804-8550 Kitakyushu, Japan,
[4]Kyushu Institute of Technology, Department of Mechanical Engineering, 804-8550 Kitakyushu, Japan,
[5]CRIEPI, Electric Power Engineering Research Laboratory, 240-0196 Yokosuka, Japan,

## ABSTRACT

We have prepared 2% Al doped ZnO (AZO) thin films on $SrTiO_3$ and $Al_2O_3$ substrates by Pulsed Laser Deposition (PLD) technique at various deposition temperatures ($T_{dep}$ = 300 °C – 600 °C). Transport and thermoelectric properties of AZO thin films were studied in low temperature range (300 K - 600 K). AZO/STO films present superior performance respect to AZO/$Al_2O_3$ films deposited at the same temperature, except for films deposited at 400 °C. Best film is the fully *c*-axis oriented AZO/STO deposited at 300 °C, with electrical conductivity 310 S/cm, Seebeck coefficient -65 μV/K and power factor 0.13 × $10^{-3}$ $Wm^{-1}K^{-2}$ at 300 K. Its performance increases with temperature. For instance, power factor is enhanced up to × $10^{-3}$ $Wm^{-1}K^{-2}$ at 600 K, surpassing the best AZO film previously reported in literature.

## INTRODUCTION

The need for energy production and conservation in the industrialized world has generated interest in effective alternative energy approaches, to overcome the dependence of mankind from traditional energy sources (carbon, oil, and fossil fuel) and reduce the $CO_2$ emission. Thermoelectric materials are considered extremely interesting from sustainable point of view because they can convert thermal energy to electrical energy [1]. The efficiency of thermoelectric energy conversion is determined by the dimensionless figure of merit

$$ZT = (\sigma.S^2).T/\kappa \qquad (1)$$

Where S: Seebeck coefficient; σ: electrical conductivity; κ thermal conductivity; T: absolute temperature [2].

Because of their poor conversion efficiency (below 20% for $Bi_2Te_3$-based devices [3]) thermoelectric materials have been restricted for a long time to a scientific scope. Efforts were made worldwide to enhance the performance of thermoelectric materials (*i.e.* the value of ZT) by increasing values of S, σ and at the same time lower the value of κ as much as possible.

In past few decades materials such as silicon-germanium alloys, metal chalcogenides, boron compounds and many more were developed for thermoelectric applications. The performances of these materials were remarkable. For example, the value of ZT for metallic thermoelectric $Bi_2Te_3/Sb_2Te_3$ multi layered films was reported up to 2.5 at T = 300 K [4]. However, their practical applications were limited because of low temperature decomposition, oxidation, vaporization or phase transition. These limitations have stimulated a lot of research on oxides as thermoelectric materials because they are thermally and electrically stable in air at high temperatures. Among oxides, ZnO ceramic has always been attracted much attention because of its versatile applications such as optical devices in ultraviolet region [5], piezoelectric transducers [6], transparent electrode for solar cells [7], gas sensors [8] and many more. ZnO *n*-type semiconductor is a potentially low-cost, nontoxic, stable thermoelectric material, that can be used up to very high temperatures because its decomposition temperature is more than 2000 °C. Pure and doped bulk ZnO has been studied as thermoelectric material for space applications, solar-thermal electrical energy production and so on [9, 10].

Thin film materials are advantageous over bulk because of light weight, quick response time, and compact size for module or sensors. For example, enhancement of thermoelectric power in Si-Ge-Au amorphous thin films [11] and $Bi_2Te_3/Sb_2Te_3$ super lattice thin film [4] has been previously reported. A limited amount of works on thermoelectric ZnO thin films [12-15] have been published so far, therefore thermoelectric ZnO thin films are not enough explored, even if they are relatively easy to prepare [16]. The main reason is the practical difficulty in the measurement of thermoelectric properties at high temperature.

In this work, we will report on 2% Al doped ZnO thin films (AZO) fabricated by Pulsed Laser Deposition (PLD) with special regard to the effect of substrate on structural, electrical and thermoelectric properties.

## EXPERIMENTAL DETAILS

The 2% Al doped ZnO thin films (AZO) thin films were grown by PLD technique using Nd:YAG laser (266 nm, 10 Hz). Pellet of $Zn_{0.98}Al_{0.02}O$ (20 mm in diameter and 3 mm in thickness) prepared by spark plasma sintering was used as target to grow the thin films. The 2 % Al was choose as best doping in the bulk [17]. Detailed description of sintered target growth is reported elsewhere [18]. The laser was shot on the dense AZO target with an energy density of about 4.2 J/cm$^2$ for deposition period of 30 min. Thin films were deposited on $SrTiO_3$ 100 (STO) and $Al_2O_3$ (100) single crystal substrates at 300 °C, 400 °C , 500 °C and 600 °C under an oxygen pressure of 200 mTorr. The target was rotated during the irradiation of laser beam. The STO substrates were glued with silver paste on Inconel plate customized for ultrahigh vacuum applications. The thickness of thin films was kept about 500 nm. Deposition parameters such as pulse frequency 10 Hz, substrate-target distance about 35 mm and rotation speed of the target 30% rpm were kept unchanged during all the deposition routes. The structural characterization was done by X-ray diffraction (XRD) (Bruker D8 Discover) and morphology was checked by scanning electron microscope (SEM) (JEOL, FESEM). The thickness and in-plane roughness were obtained by 3D-microscope Keyence VK-9700. The electrical conductivity vs temperature ($\sigma$-T) characteristics were measured by conventional four-probe technique from 300 to 600 K with hand-made apparatus consisting in a current source (ADCMT 6144), temperature controller (Cryo-con 32) and nano voltmeter (Keithley 2182A). Seebeck coefficient was measured by commercially available system (MMR Technologies Inc.) in the temperature range 300 K - 600 K. Carrier concentration and mobility at room temperature were evaluated by means of Quantum Design PPMS.

## RESULTS

**Figure 1.** TEM cross-sectional images of AZO thin films deposited on: (a) $SrTiO_3$ and (b) AZO/$Al_2O_3$ single crystal substrates

Figure 1 (a, b) shows cross sectional TEM image of AZO thin film deposited at 500 °C on STO and $Al_2O_3$ substrates. Grains are highly connected and vertically aligned, showing a typical columnar growth. The sample deposited on STO presents several *a*-axis oriented grains, while the sample deposited on $Al_2O_3$ is fully *c*-axis oriented. It is not possible to find a clear dependence of the grain size and morphology with deposition temperature: according to SEM observations (not reported here), for both series of samples morphology evolves from big rounded grains (diameter 100-200 nm), to small elongated grains (length ~ 30-100 nm, at 400 °C), then big elongated grains (length ~ 100-300 nm, at 500 °C) and eventually hexagonal grains (side 100-200 nm, at 600 °C). Furthermore, AZO films deposited on STO at 400 °C and 600 °C show several pores, while the films fabricated at 300 °C and 500 °C are well connected. In case of films deposited on $Al_2O_3$, pores are always absent and the grains are highly connected for all deposition temperature.

XRD $\theta$–$2\theta$ scans (Fig. 2 a, b) of the films show only peaks belonging to wurtzite hexagonal structure. Secondary phases such as $ZnAl_2O_4$ did not form during thin film deposition process. The films grown on STO at 300 °C and 400 °C show only (002) peak indicating full *c*-axis oriented growth. Additional (110) peak, corresponding to *a*-axis orientation, appears increasing the deposition temperature to 500 °C and 600 °C. Indeed, the sample deposited at 600 °C presents (100) peak as well. All the films deposited on $Al_2O_3$ are *c*-axis oriented.
For both AZO/STO and AZO/$Al_2O_3$ samples, the intensity of (002) peak, *i.e.* the crystallinity of films, decreases with increasing deposition temperature.

The $\phi$-scans of thin films are also shown in figure 3 (a, b). On STO the films deposited at 400 °C and 600 °C show 12 peaks instead of the expected 6, indicating a mixed structure with 30° rotated and not-rotated hexagonal AZO unit cells oriented along *c*-axis of STO. The other two films have no sign of epitaxial growth. Lattice mismatch of AZO on STO can be calculated

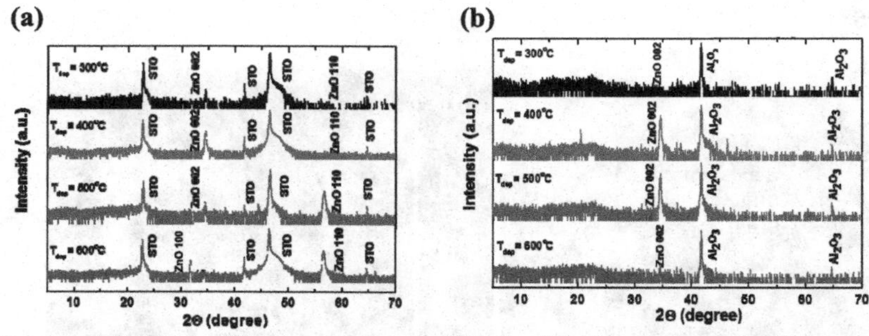


**Figure 2.** θ–2θ X-ray diffraction patterns of AZO thin films deposited on (a) $SrTiO_3$ and (b) $Al_2O_3$ single crystal substrates

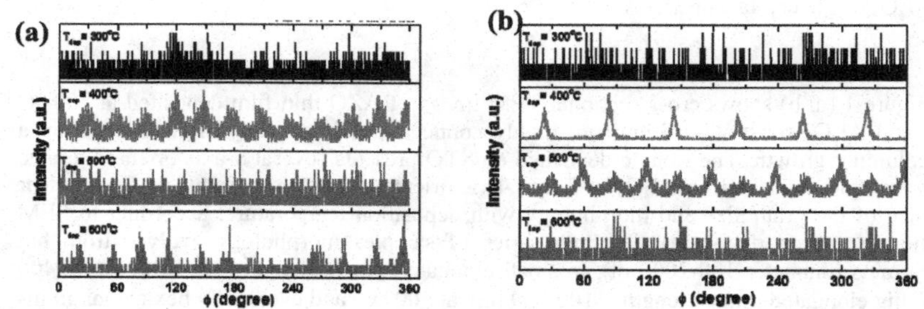


**Figure 3.** ϕ–scan of AZO thin films deposited on (a) $SrTiO_3$ and (b) $Al_2O_3$ single crystal substrates

as $\varepsilon_{c,STO} = -2\%$. For *a*-axis orientation, 4 peaks in the ϕ scan are expected, with a lattice mismatch $\varepsilon_{a,STO} = -6\%$. On $Al_2O_3$ substrate, films deposited at 400 °C show 6-fold symmetry with the 6 peaks shifted by 30° respects to the peaks of $Al_2O_3$ substrate (not shown here). This rotation allows to have quite large epitaxial strain $\varepsilon_{c,Al2O3} = 19\%$. Increasing $T_{dep}$, the alignment of AZO on $Al_2O_3$ becomes poor, with decreasing intensity of the peaks in the ϕ scan. The film deposited at 300 °C also show poor epitaxy, consistently with low intensity of peaks in the θ-2θ pattern. In film deposited at 500 °C 6 additional peaks are present, indicating a mixed structure with 30° rotated and not-rotated hexagonal unit cells on $Al_2O_3$. For both AZO/STO and AZO/$Al_2O_3$ thin films, despite to large values of ε, there is no effect on crystalline parameters, which remain almost unchanged respect to the bulk values (Table I).

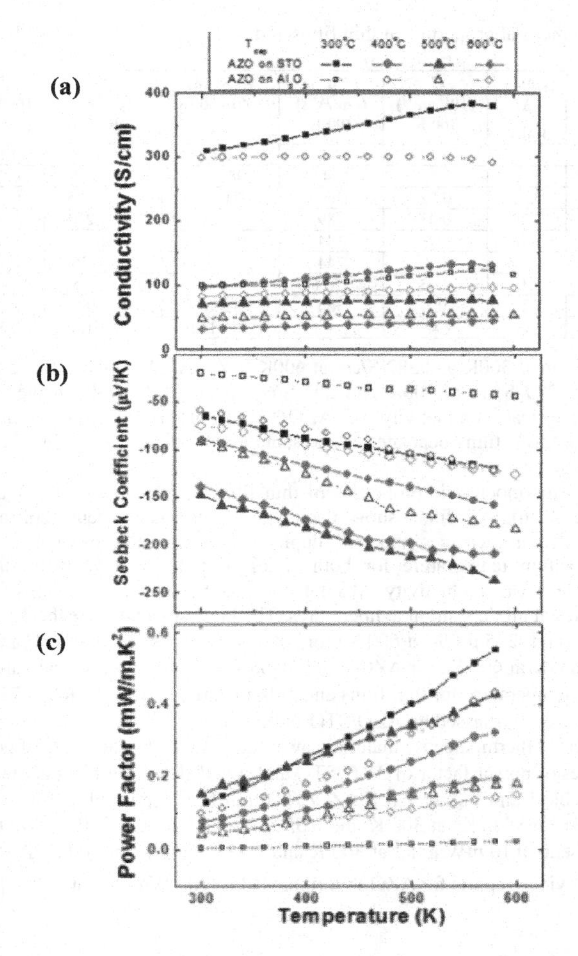


**Figure 4.** Temperature dependence of transport and thermoelectric properties: (a) electrical conductivity, (b) Seebeck coefficient and (c) power factor of AZO/STO (filled symbols)and $AZO/Al_2O_3$ (empty symbols) films.

Transport and thermoelectric properties of AZO thin films are summarized in Table I.

Figure 4 (a) shows electrical conductivity *vs.* temperature (σ-T) for AZO thin films deposited on STO and $Al_2O_3$. σ increases with temperature, with a typical semiconducting behavior. AZO/ STO film deposited at 300 °C shows the larger σ in the whole range of

**Table I.** Electrical and thermal properties of thin films and bulk pellet of AZO at 300K/600K

| Sub. | $T_{dep}$ (°C) | a axis (Å) 300 K | c axis (Å) 300 K | C ($10^{19}$ $cm^{-3}$) 300 K | μ ($cm^2$/V s) 300 K | σ (S/cm) 300K/600K | S (μV/K) 300K/600K | PF (mW/m $K^2$) 300K/600K |
|---|---|---|---|---|---|---|---|---|
| $SrTiO_3$ | 300 | - | 5.21 | 0.12 | 1653 | 310/382 | -65/-121 | 0.13/0.55 |
| | 400 | - | 5.20 | 2.55 | 24 | 98/133 | -90/-163 | 0.08/0.32 |
| | 500 | 3.25 | 5.21 | 0.02 | 260 | 71/77 | -151/-245 | 0.15/0.43 |
| | 600 | 3.25 | 5.20 | 0.03 | 789 | 32/45 | -138/-214 | 0.06/0.19 |
| $Al_2O_3$ | 300 | - | 5.18 | 0.5 | 134 | 99/116 | -20/-44 | 0.004/0.02 |
| | 400 | - | 5.19 | 0.3 | 744 | 299/291 | -58/-126 | 0.10/0.43 |
| | 500 | - | 5.19 | 0.03 | 1541 | 49/54 | -74/-125 | 0.04/0.18 |
| | 600 | - | 5.21 | 0.1 | 366 | 82/96 | -92/-183 | 0.05/0.15 |
| Bulk [18, 19] | | 3.25 | 5.20 | - | - | 206/152 | -132/-150 | 0.35/0.34 |

temperatures: 310 S/cm at 300K and 382 S/cm at 600K. AZO/$Al_2O_3$ film deposited at 400 °C is clearly the best with 299 S/cm at 300K and 291 S/cm at 600K. Overall, in the whole range 300~600 K, the electrical conductivity of AZO/STO films is superior to the electrical conductivity of AZO/$Al_2O_3$ films deposited at the same temperature, except for films deposited at 400 °C and 600 °C.

We measured thermoelectric properties of thin films at 300 K and 600 K as shown in Table I and Figure 4 (b). All films show negative Seebeck coefficient, indicating *n*-type conduction due to oxygen vacancies and $Al^{3+}$ doping. Absolute value of Seebeck coefficient increases with deposition temperature for both series of thin films, showing the opposite behavior respect to electrical conductivity. Absolute value of Seebeck coefficient is found higher for AZO/STO samples at all deposition temperatures. The highest value of Seebeck coefficient is -151 μV/K at 300 K (and -245 μV/K at 600 K) for AZO/STO deposited at 500 °C and -92 μV/K at 300 K (and -183 μV/K at 600 K) for AZO/$Al_2O_3$ deposited at 600 °C. Typical values reported in the same range of temperature for thin films and bulk materials (-150 μV/K at 673 K [9], -110 μV/K at 650 K [20]) were surpassed by AZO/STO films.

The efficiency of thermoelectric material can be determined by the value of power factor, PF = $\sigma.S^2$. The values of power factor of AZO/STO and AZO/$Al_2O_3$ thin films at 300 K and 600 K are reported in Table I and Figure 4 (c). AZO/STO thin film deposited at 300 °C shows the best power factor: 0.13 mW/m.$K^2$ at 300 K and 0.55 mW/m.$K^2$ at 600 K. Best AZO/$Al_2O_3$ film ($T_{dep}$ = 400 °C) presents 0.10 mW/m.$K^2$ at 300 K and 0.43 mW/m.$K^2$ at 600 K. Performance of both films exceed previous reports for AZO thin films: 0.35 $\times 10^{-3}$ $Wm^{-1}K^{-2}$ at 740 K [14].

## DISCUSSION

Two points require explanation: why in both series one sample presents superior electrical conductivity respect to the others, and why films deposited on STO have generally larger PF than films deposited on $Al_2O_3$.

Focusing on AZO/STO films at first, the film deposited at 300 °C shows the highest electrical conductivity at room T (about 310 S/cm) while the film deposited at 600 °C shows electrical conductivity of 32 S/cm at room temperature, lower than the bulk material (209 S/cm [18, 19]). The trend of electrical conductivity cannot be simply related to the microstructure, otherwise the samples deposited at 300 °C and 500 °C, which present similar amount of large grains (and consequently small amount of grain boundaries) should have similar values of $\sigma$. Indeed, as reported in Table I, there is not direct proportionality between carrier concentration and electrical conductivity: for example, the sample with the highest carrier concentration ($T_{dep}$ = 400 °C) is the second sample in terms of $\sigma$. Furthermore, carrier concentrations of our films are far from the value $10^{20}$ $cm^{-3}$, reported as optimal for thermoelectric properties [21]: this is the consequence of the oxygen atmosphere used in the deposition. Overall, to explain the electrical conductivity behavior of the AZO/STO films it is possible to invoke the anisotropy of $\sigma$ with the crystallographic orientation: fully *c*-axis oriented samples (deposited at 300 and 400 °C) possess the highest electrical conductivity, followed by sample with *c* and *a* axis orientation (500 °C) and eventually by polycrystalline sample deposited ant 600 °C. This reminds the scenario reported by Abutaha *et al* [14] for AZO films deposited on $LaAlO_3$ single crystals at T = 573-1273 °C. The main difference with our AZO-STO samples is their extremely large $\sigma$ (941 S//cm at 300K for best *c*-axis oriented film) and the presence of fully *a*-axis oriented films.

For AZO/$Al_2O_3$ films, the grain size and carrier concentration explanations fail for the same reasons discussed above. The *c*-axis/ *a*-axis interplay argument cannot be used since all the films are *c*-axis oriented. It is only possible to invoke the superior crystallinity of the sample deposited at 400 °C to justify his larger $\sigma$.

In order to explain why AZO/STO films are superior to AZO/$Al_2O_3$ films, we can calculate the concentration of dislocations by

$$N = [(a_f - a_s)/a_{avg}^2]^2 \quad (2)$$

where $a_f$, $a_s$ and $a_{avg}$ are the thin film, substrate and average lattice parameter, respectively [22]. N is ~ $10^{10}$ $cm^{-2}$ for AZO/STO film and ~ $10^{11}$ $cm^{-2}$ for AZO/ $Al_2O_3$. These values are consistent with the values of epitaxial strain reported earlier in this paper. $\sigma$ is reported to decrease and S to increase with N [23]: this consideration may explain the different performance of the two films. In order to reach optimal N = $10^6$~$10^8$ $cm^{-2}$ [23] and further improve the thermoelectric performance, it is necessary to release the epitaxial strain by growing thicker films [24] (1000 nm or more) or inserting buffer layers [25].

## SUMMARY

We prepared 2% Al-doped ZnO (AZO) thin films by Pulsed Laser Deposition (PLD) on $SrTiO_3$ (STO) and $Al_2O_3$ single crystal substrates at $T_{dep}$ = 300 °C ~600 °C. Films deposited on $Al_2O_3$ are epitaxial and fully *c*-axis oriented showing only (002) peak in q-2q XRD patterns, independently of $T_{dep}$. Films deposited on STO are c-axis oriented for $T_{dep} \leq 400$°C, though at higher $T_{dep}$ *a*-axis orientation (110 peak) also appears. Except for $T_{dep}$ = 400 °C, films deposited on STO always shows higher values of S and $\sigma S^2$ (power factor) in comparison with films deposited on $Al_2O_3$. Best values (T = 600 K) are obtained on STO at $T_{dep}$ = 300 °C: $\sigma_{STO}$ = 382 S/cm, $S_{STO}$ = -121 µV/K, $(\sigma S^2)_{STO}$ = 0.55 mW/m.$K^2$ overcoming the best results reported in literature. Different behavior is explained by larger epitaxial strain ($\varepsilon_{c,Al2O3}$ ~ 15%) and larger concentration of dislocations (N ~ $10^{11}$ $cm^{-2}$) of AZO/$Al_2O_3$, while $\varepsilon_{c,STO}$ ~ 6%, $\varepsilon_{a,STO}$ ~ 2% and N ~ $10^{10}$ $cm^{-2}$ for AZO/STO. Further improvement in thermoelectric performance is expected by release of epitaxial strain (thicker films or insertion of buffer layers).

## REFERENCES


1. T. J. Seebeck, *Abh. Akad. Wiss. Berlin*, **289** (1822).
2. H. Böttner, *Mater. Res. Soc. Symp. Proc.* **N01-01**, 1166 (2009).
3. D. M. Rowe, *Thermoelectrics Handbook: Macro to Nano.* (Boca Raton: CRC/Taylor & Francis, 2006).
4. R. Venkatasubramanian, E. Siivola, T. Colpitts, B. O'Quinn, *Nature* **413**, 517 (2001).
5. G. M. Ali, P. Chakrabarti, *J. Phys. D: Appl. Phys.* **43**, 415103 (2010).
6. P. X. Gao and Z. L. Wanga, *J. Appl. Phys.* **97**, 044304 (2005).
7. M. Law, L. E. Greene, J. C. Johnson, R. Saykally, P. Yang, *Nat. Mater.* **4** , 455 (2005).
8. M. Kaur, S. V. S. Chauhan, S. Sinha, M. Bharti, R. Mohan, S. K. Gupta, J. V. Yakhmi: *J. Nanosci. Nanotech.* **9**, 5293 (2009).
9. M. Ohtaki, T. Tsubota, K. Eguchi, and H. Arai, *J. Appl. Phys.* **79**, 1816 (1996).
10. J. P. Wiff, Y. Kinemuchi, and K. Watari, *Mater. Lett.* **63**, 2470 (2009).
11. H. Uchino, Y. Okamoto, T. Kawahara and J. Morimoto, *Jpn. J. Appl. Phys.* **39**, 1675 (2000).
12. Y. Inoue, M. Okamoto, T. Kawahara, Y. Okamoto and J. Morimoto, *Materials Transactions* **46-7**, 1470 (2005).
13. K. P. Ong, D. J. Singh, and P. Wu, *Phys. Rev.* B **83**, 115110 (2011).
14. A.I. Abutaha, S, R. Sarath Kumar, and H. N. Alshareef, *Appl. Phys. Lett.* **102**, 053507 (2013).
15. N. Vogel-Schäuble, Y. E. Romanyuk, S. Yoon, K. J. Saji, S. Popuolh, S. Pokrant, M. H Aguirre, A. Wiedenkaff, *Thin Sol. Films* **520** , 6869 (2012).
16. B. Singh, Z. A. Khan, I. Khan, and S. Ghosh, *Appl. Phys. Lett.* **97**, 241903 (2010).
17. P. Mele, K. Matsumoto, T. Azuma, K. Kamesawa, S. Tanaka, J. Kurosaki, K. Miyazaki: *Mater. Res. Soc. Symp. Proc.* **1166**, 3 (2009).
18. P. Mele: submitted to *Mat. Sci. Eng. B*, (2013).
19. For bulk AZO, the value of S, $\sigma$ and Power Factor at 300K are taken by extrapolation of S and $\sigma$ from Mele et al. [18]
20. H. Hiramatsu, H. Ohta, W.-S. Seo and K. Koumoto, *J. Jpn. Soc. Powder and Powder Metall.* **44**, 44 (1997).
21. A. Ioffe: *Semiconductor Thermoelements and Thermoelectric Cooling* (Infosearch Ltd., London, 1957).
22. J. M. Woodall, G. D. Pettit, T. N. Jackson, C. Lanza, K. L. Kavanagh, J. W. Mayer, *Phys. Rev. Lett.* **51**, 1783 (1983).
23. J. R Waiting and D. J. Paul, *J. Appl. Phys.* **110**, 114508 (2011).
24. S. H. Park, T. Hanada, D. C. Oh, T. Minegishi, H. Goto, G. Fujimoto, J. S. Park, I. H. Im, J. H. Chang, M. W. Cho, T. Yao, *Appl. Phys. Lett.* **91**, 231904 (2007).
25. T. Nakamura, Y. Yamada, T. Kusumori, H. Minoura, H. Muto, *Thin Sol. Films* **411**, 60 (2002).

Mater. Res. Soc. Symp. Proc. Vol. 1543 © 2013 Materials Research Society
DOI: 10.1557/opl.2013.672

**Mapping thermal resistance around vacancy defects in graphite**

Laura de Sousa Oliveira and P. Alex Greaney
School of Mechanical, Industrial, & Manufacturing Engineering
Oregon State University, Corvallis, OR 97331

# ABSTRACT

High purity bulk graphite is applicable in many capacities in the nuclear industry. The thermal conductivity of graphite has been found to vary as a function of how its morphology changes on the nanoscale, and the type and number of defects present. We compute thermal conductivities at the nanolevel using large scale classical molecular dynamics simulations and by employing the Green-Kubo method in a set of *in silico* experiments geared towards understanding the impact of defects in the thermal conductivity of graphite. We present the results obtained for systems with 1–3 vacancies, and compile a summary of some of the methods applied and difficulties encountered.

# INTRODUCTION

Graphite is applicable in many capacities in the nuclear industry. It is used in gaskets, sealants, and liners, but most importantly, it is used as a moderator and a reflector, and its unique properties are being exploited in order to develop high-tech fuel elements for next-generation nuclear reactors. While graphite has been comprehensively studied since the 1950s [4], there are aspects of its thermal conductivity ($\kappa$) which have yet to be well understood. Graphite is highly anisotropic and the thermal conductivity along the basal plane ($\kappa_a$) differs significantly from that along the c-axis ($\kappa_c$), with an experimentally computed anisotropy ratio ($\kappa_a/\kappa_c$) just below 210 at 300K in near-ideal graphite [3]. Furthermore, the thermal conductivity in bulk graphite varies as a function of how the material is manufactured and its exposure to radiation and high temperatures within a reactor. Our motivation is to establish a systematic understanding of how defect type, number and different defect-type ensembles affect thermal resistance in graphite. Defects can occur at different scales, and while grain boundaries, porosity and amorphous regions, for instance, can significantly affect thermal transport, in our first steps towards achieving our goal we examine point defects. In this Proceedings paper we report on a collection of vacancy defects. In addition, we summarize the methods we have developed for studying phonon scattering around defects and some of the difficulties that arise when computing $\kappa$ in graphite.

# METHODS

Simulations were performed using large-scale equilibrium classical molecular dynamics. More specifically, we use the LAMMPS [6] software, distributed by Sandia National Laboratories. Molecular dynamics (MD) is a powerful tool for understanding thermal behavior and phonon scattering. However, there are several limitations to MD that make qualitative predictions of thermal conductivity unlikely. To mitigate this and to gain insight into how thermal resistance varies at the atomic-level and as a function of the different defects, we perform a comparative analysis. In this study, we compare the thermal conductivity and corresponding anisotropy ratios obtained between the different defective systems and the perfect crystal, as well as in different regions within each system. $\kappa$ is computed along each axis, for each system and in specific regions within each system, by employing the AIREBO [7] [1] potential, as it is implemented in the LAMMPS MD package, and the Green-Kubo formulation [2] [5] in the microcanonical ensemble (NVE). The Green-Kubo method (see Equation 1), derived from the fluctuation-dissipation theorem, utilizes the ensemble average of the autocorrelation of the instantaneous heat flux (J) to compute $\kappa$ in a given direction. In Equation 1, T is the temperature in the system or region considered, V is the corresponding volume, and $k_B$ is Boltzmann's constant.

$$\kappa_x = \frac{V}{k_B T^2} \int_0^\infty \langle J_x(t) J_x(t+\Delta t) \rangle \, dt \tag{1}$$

To map the local thermal conductivity we divide the compute cell into a system of 550 overlapping sub-regions (5, 10, and 11 along the x, y and z directions respectively). J is computed in each sub-region, and its autocorrelation function integrated to produce a local $\kappa$. The regional $\kappa$ values obtained are used to compute iso-surface maps, which are discussed in the Results section. The perfect crystalline system simulated consists of 10648 atoms, approximately $(265.6 \times 459.6 \times 758.2)$ nm$^3$, from which 1–3 atoms are removed along the basal plane, in a centrally located region of the system (see Fig. 1). The difference in the size of the system along the x and y directions results in different thermal conductivities along these axes and allows us to identify size artifacts.

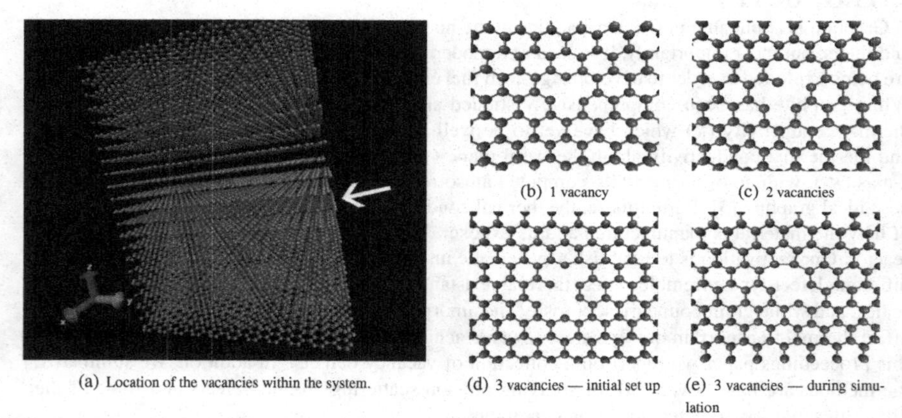

(a) Location of the vacancies within the system. (b) 1 vacancy (c) 2 vacancies (d) 3 vacancies — initial set up (e) 3 vacancies — during simulation

**Figure 1:** In (a), the arrow points to where both the single vacancy and the 2–3 in-plane vacancy clusters are placed within each system. This image corresponds to the cluster with 3 vacancies. (b) and (c) illustrate the single vacancy and double vacancy cluster within the plane, respectively. (d) shows where the three vacancies are originally introduced, and (e) shows how the atoms rearrange themselves during the NVT simulation.

The perfect and defective systems are simulated at 300 K* with a time step of $2 \times 10^{-7}$ ns for a total duration of 0.6 ns. The autocorrelation is computed for 0.015 ns. The isothermal-isobaric ensemble (NPT) is used to bring the systems to 300 K, at which point we switch to NVE and

*It is important to note that while this is well below the Debye temperature, our goal is a comparative analysis of phonon scattering from and around the defect — not phonon-scattering.

simulate them for different lengths of time in order to obtain different initial atomic arrangements for each system. Overall, our results are averaged over 20 separate simulations for the perfect crystalline system, 11 simulations for the 2–vacancy-cluster, and 10 simulations for the remaining defective systems.

## FEATURES OF THE AUTOCORRELATION FUNCTION

When computing the autocorrelation, it was observed that rare events take place which make a rapid and large change to the total heat current autocorrelation function $(\langle J_X(t)J_X(t+\Delta t)\rangle)$. These fluctuations require a long averaging time to diminish and are more prevalent in the heat current along the basal plane than along the c-axis. The rare events, examples of which can be seen in Figs. 2(a), (b) and (c) are not present in every run, but when they occur they are large enough to affect the resulting $\kappa$. To obtain a statistically significant sample of runs which would include them, we simulate multiple system sizes for the perfect crystalline graphite and, using the same method described in the Methods section, produce several runs for each system size. It is possible that these events are artifacts of the system size; however, we do not notice any reduction in the number of these events with increasing system size. Alternatively, the rare fluctuations observed could be a manifestation of the Fermi–Pasta–Ulam–Tsingou problem. The effect these rare fluctuations have on the thermal conductivity is best illustrated in Figs. 2(d), (e) and (f), which shows $\kappa$ computed as a simulation progresses. In other words, for each autocorrelation plotted as in Figs. 2(a), (b) and (c) there is a corresponding $\kappa$ value obtained with the Green-Kubo formulation by integrating along the "Autocorrelation Time" axis — these are plotted for many simulations in Figs. 2(d), (e) and (f). While it is clear that $\kappa$ tends to converge, we can observe spikes in its value along some runs, which may occur at any time within a given run, even as it appears to be converging, and which lasts for varying periods of time. In response to these events, several protocols for computing thermal conductivities using different integrating schemes (see Fig. 3) were tested. Specifically, we truncated the autocorrelation function at different times, and tried integrating multiple exponential fits to the autocorrelation (only the best fit obtained is actually shown in Fig. 3). Integrating the autocorrelation for 15 ps yields systematically higher thermal conductivities, but larger error bars. Since the rare events do not take place in every run and affect each run differently, integrating over the tail of the autocorrelation function increases the variability of the resulting $\kappa$ values, resulting in larger error bars. Integrating for a shorter amount of time will exclude the tail and result in consistently smaller values of $\kappa$, but does reduce the error bars. Ultimately we are constrained by computer time. Our goal is to gather as good statistics — including rare events — as possible that allow a meaningful comparison of predicted $\kappa$ values that are not distorted by a low number of events. To this end, we are willing to underestimate kappa in return for performing more, shorter simulations to improve ergodic sampling. Provided that we adhere to the same integration protocol, we should be able to derive significant insight from the various defective systems simulations concerning how the thermal resistance in graphite is affected by point defects.

While LAMMPS may be used to compute the autocorrelation function for the entire system, to compute maps for how the thermal conductivities vary within each system, we use our own code to compute the autocorrelations. The code was test for agreement with LAMMPS, as can be seen in Fig. 3. The following results are computed using our code.

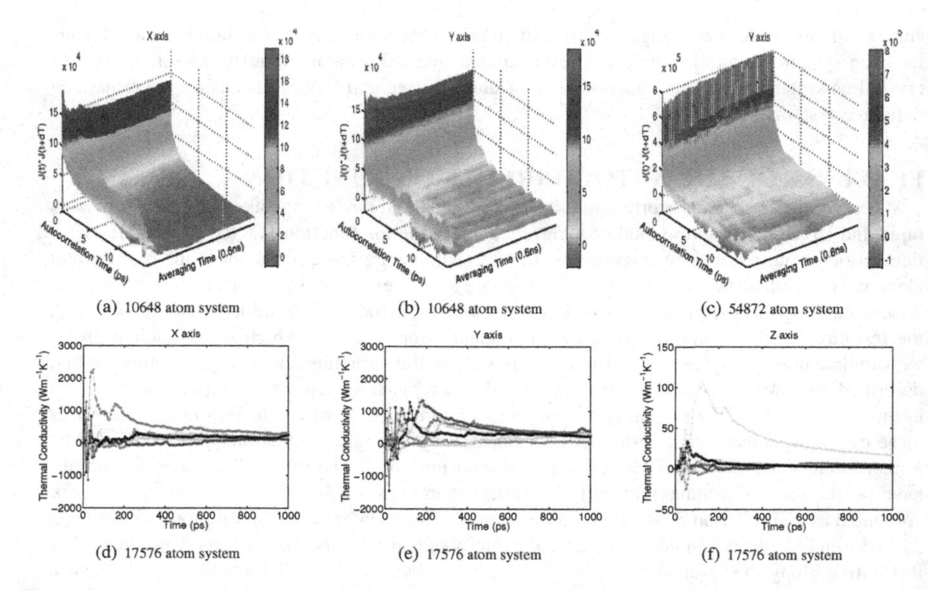

(a) 10648 atom system (b) 10648 atom system (c) 54872 atom system

(d) 17576 atom system (e) 17576 atom system (f) 17576 atom system

**Figure 2:** In the first horizontal block of figures, the autocorrelation has been plotted multiple times as it is computed (and thus averaged) as the simulation progresses. Initially, there are few values that contribute towards the ensemble average and thus the initial autocorrelation functions are noisy. As the averaging time progresses the autocorrelations become smoother with the exception of well-defined crests and troughs (aforementioned as rare events) which fade slowly, or not at all during the total simulation time. (a) and (b) correspond to the same run in x and y, respectively. (c) shows the autocorrelations of the instantaneous heat-flux computed over time in the x direction for a perfectly crystalline system with 54872 atoms. The second block of figures ((d)–(f)) corresponds to a 17576 atom system and each curve color corresponds to a a single run. Each plot corresponds to x, y and z values. The thermal conductivity values are obtained using the autocorrelation function computed at different run times for each run within each system.

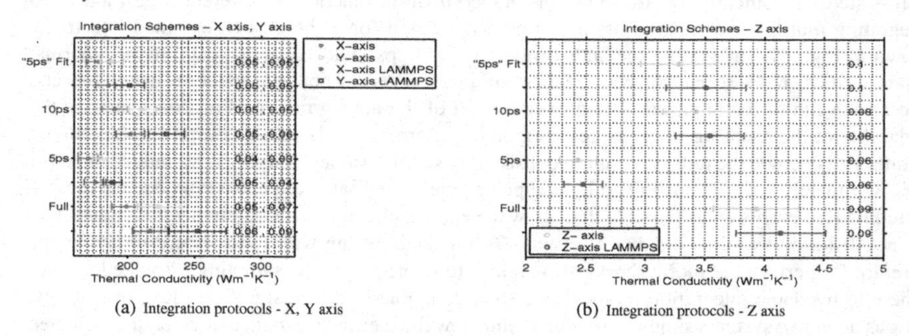

(a) Integration protocols - X, Y axis (b) Integration protocols - Z axis

**Figure 3:** While these plots correspond to the 10648 atom system, the change in the error bars and thermal conductivity values with respect to the integrating scheme is systematically the same for all systems. The numbers on the right of each plot correspond to the relative standard error (not normalized out of 100%) for the x and y-axis, respectively, in (a) and for the z-axis, in (b). The first values for each protocol correspond to the autocorrelation function computed with our code.)

# RESULTS

The thermal conductivity values obtained in each axis for the different vacancy cluster defects and the perfect system can be found in Fig. 4. While there is no statistically significant change in

$\kappa$ along the c-xis, we observe significant differences in the $\kappa_x$ and $\kappa_y$ values between the perfect system and the 2 and 3–vacancy-clusters. In these clusters we see the thermal conductivity go up with respect to the perfect crystal and the single-vacancy defect, and in the 3–vacancy-cluster case, we observe a large difference between $\kappa_x$ and $\kappa_y$. During the simulations, the vacancy and the vacancy-clusters diffuse along the plane they were inserted in. The anisotropy ratios (Fig. 4(c)) reflect the combined changes in the thermal conductivities along the basal plane and along the c-axis.

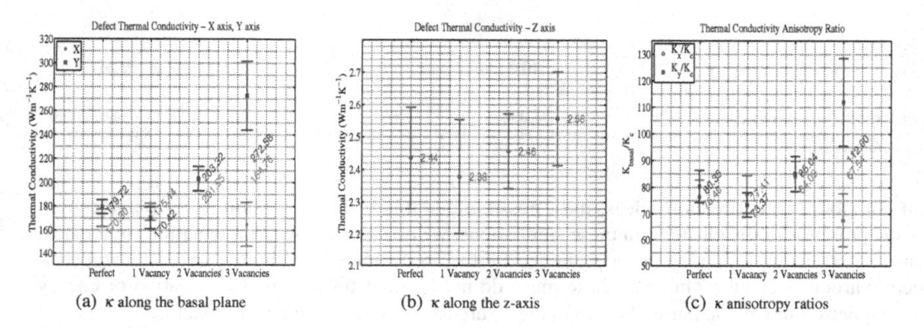


(a) $\kappa$ along the basal plane (b) $\kappa$ along the z-axis (c) $\kappa$ anisotropy ratios

**Figure 4:** (a) and (b) are the $\kappa$ values obtained for the perfect, the single-vacancy and 2 and 3–vacancy-cluster systems, both along the basal plane ((a)) and along the c-axis ((b)). (c) is the anisotropy ratio computed for both x and y.

The following $\kappa$-maps (Fig. 5) were intended to shed some light on how the thermal conductivities change around the defects. These maps were obtained by plotting the iso-surfaces of $\kappa$ using regionally computed thermal conductivities, i.e., by integrating the local J correlation function. These maps are very noisy and show no systematic local change in $\kappa$. This could be possibly due to the diffusion of the vacancy and the vacancy-clusters. While it is interesting that the noise is stratified and that the fluctuations in thermal conductivity between regions are large, there doesn't seem to be a clear distinction between how thermal conductivities propagate and what their values are in the regions near the defects. Table 1 contains the ratios between the maximum and minimum thermal conductivity values found within each system, and there appears to be no significant difference between the maximum and minimum regional $\kappa$s between systems.

| | Perfect System | 1 Vacancy | 2 Vacancies | 3 Vacancies |
|---|---|---|---|---|
| $Max(\kappa_x)/Min(\kappa_x)$ | 5.1987 | 7.18629 | 6.48254 | 9.09487 |
| $Max(\kappa_y)/Min(\kappa_y)$ | 7.71476 | 9.85168 | 12.0638 | 13.6437 |
| $Max(\kappa_z)/Min(\kappa_z)$ | 27.0945 | 24.7075 | 7.61275 | 10.6363 |

**Table 1:** Ratios between the maximum and minimum mean regional thermal conductivities within each system.

## CONCLUSION

We have computed thermal conductivities of graphite containing a collection of vacancy defects. In doing so, we have found that the vacancy-clusters increase the thermal conductivity in-plane,

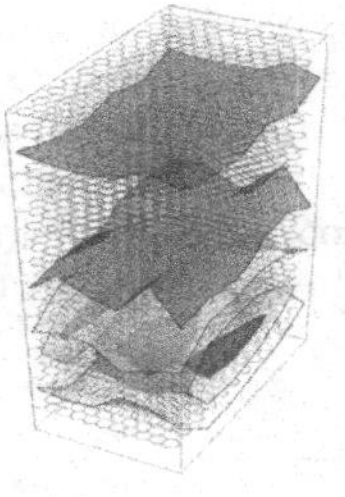

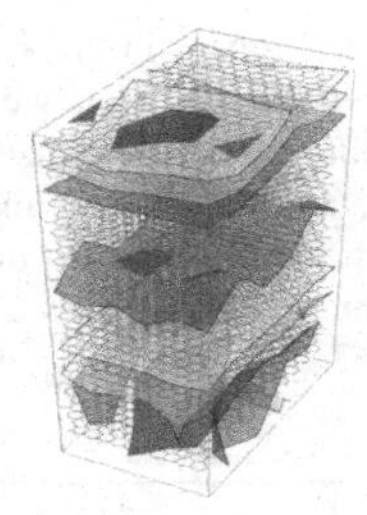

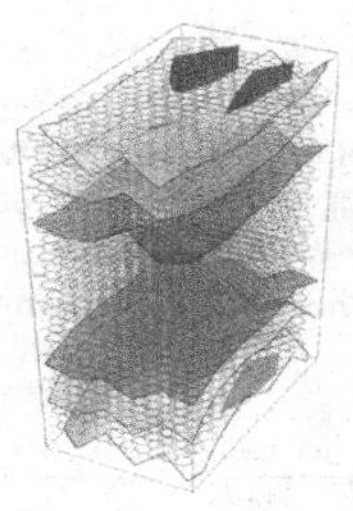

(a) Perfect - Y-axis (b) Perfect - Z-axis (c) 3 Vacancies - Y-axis

**Figure 5:** Thermal conductivity maps. (a) and (b) correspond to the y and z (in this order) thermal conductivity iso-surfaces obtained for a perfect system. (c) corresponds to the 3–vacancy-cluster, in the y direction. These maps were computed from the mean value of $\kappa$ obtained for 550 regions for each system.

but not along the c direction – it is, thus, possible that this is due to tension in the plane caused by the presence of the clusters. In these simulations, we observe how large fluctuations in the heat current correlation function pose a challenge for the prediction of $\kappa$. Finally, we have mapped $\kappa$ locally around defects. Thus far, these maps do not account for the cross correlation of energy current between regions, but will do so in the future as we continue to develop them.

## ACKNOWLEDGMENTS

We thank Phillip Jenks and Rodrigo Lopes for their assistance with some of the C++ code. This work used the Extreme Science and Engineering Discovery Environment (XSEDE), which is supported by the National Science Foundation grant number OCI-1053575.

## REFERENCES

[1] Harrison, Stuart, Ni, Sinnott, Brenner, Shenderova. *J. Phys: Cond. Matt.*, 14:783–802, 2002.

[2] Melville S. Green. time-dependent phenomena. ii. irreversible processes in fluids. *The Journal of Chemical Physics*, 22(3):398–413, 1954.

[3] CN Hooker, AR Ubbelohde, and DA Young. *Proceedings of the Royal Society of London. Series A. Mathematical and physical sciences*, 284(1396):17–31, 1965.

[4] B.T. Kelly. *Physics of Graphite*. RES mechanica monographs. Kluwer Academic Pub, 1981.

[5] Ryogo Kubo. theory and simple applications to magnetic and conduction problems. *Journal of the Physical Society of Japan*, 12(6):570–586, 1957.

[6] Steve Plimpton et al. *Journal of Computational Physics*, 117(1):1–19, 1995.

[7] Harrison Stuart, Tutein. *J. Chem. Phys*, 112:6472–6486, 2000.

Mater. Res. Soc. Symp. Proc. Vol. 1543 © 2013 Materials Research Society
DOI: 10.1557/opl.2013.671

# Thermal conductivity of regularly spaced amorphous/crystalline silicon superlattices. A molecular dynamics study

Konstantinos TERMENTZIDIS[1,*], Arthur FRANCE-LANORD[1], Etienne BLANDRE[1], Tristan ALBARET[2], Samy MERABIA[2], Valentin JEAN[1] and David LACROIX[1]
[1]Université de Lorraine, LEMTA, CNRS UMR 7563, BP 70239, Vandœuvre les Nancy cedex, France
[2] Université de Lyon-1, ILM, CNRS UMR 5306, Bâtiment Kastler, 10 rue Ada Byron, 69622 Villeurbanne, France.
*konstantinos.termentzidis@univ-lorraine.fr

## ABSTRACT

The thermal transport in amorphous/crystalline silicon superlattices with means of molecular dynamics is presented in the current study. The procedure used to build such structures is discussed. Then, thermal conductivity of various samples is studied as a function of the periodicity of regular superlattices and of the applied temperature. Preliminarily results show that for regular amorphous/crystalline superlattices, the amorphous regions control the heat transfer within the structures. Secondly, in the studied cases thermal conductivity weakly varies with the temperature. This, points out the presence of a majority of non-propagating vibrational modes in such systems.

## INTRODUCTION

The amorphous/crystalline superlattices (a/c SLs) with large conduction band discontinuities can be used for resonant-tunnelling diodes, modulation-doped field-effect transistors and quantum-well infrared photo-detectors[1]. They are also interesting candidates for low-cost thermoelectric power devices[2]. These SLs can be made of materials which display large lattice mismatch. Furthermore, they can have interfaces which are essentially defect-free and atomically sharp[3].

To the best of our knowledge the only investigation of the heat transfer through a/c SLs interfaces was made by Von Alfthan[4] et al. They conclude that the thermal conductivity of such superlattices is mainly ruled by the presence of amorphous regions, whatever are their sizes. We extend this study with the prediction of the Kapitza resistance and we study the effect of the SLs parameters such as period and temperature. A recent theoretical work of Donadio and Galli on crystalline silicon nanowires with amorphous surfaces showed that $k$ is not affected by the temperature and it is close to the one of amorphous materials[5]. This unusual temperature dependence is explained by the presence of a majority of non-propagating vibrational modes.

There is a lack of information about a/c silicon interfaces, especially in what concerns both SLs and nanowires. In our study SLs of a-Si/c-Si preliminary results will present the thermal conductivity appraisal as a function of the SL's periodicity. Then, the mean temperature of structures will be changed in order to assess k variations. Hereafter follows the simulation method description and the obtained results discussion. Eventually, conclusions and perspectives to this work are provided.

## SIMULATION METHOD

In this study two molecular dynamics codes were used. The first one is a homemade molecular dynamics code, which is used to build amorphous silicon regions. The latter one was chosen in order to obtain an accurate description of the amorphous silicon bulk state and to relax the combination of amorphous/crystalline regions. The purpose being to built amorphous/crystalline superlattices with high quality interfaces and totally reproducible structures. The control of the quality of interfaces is based first on the atomic potential energies of all atoms in the direction perpendicular to the interfaces, and secondly on the radial distribution function or the pair distribution function (PDF). The first rule guarantees that atoms belonging to both crystalline and amorphous regions, have energies close to the ones of atoms of the bulk state (amorphous or crystalline). The second rule states that atoms in the amorphous region must follow a radial ordering distribution function which gives the probability to find an atom at a given distance r from another atom.

Figure-1 shows a couple of a-Si/c-Si SLs obtained by the technique described in the article of Fusco et al[6] with a superlattice's periodicity of $24a_0$ with 6 repetitions (up) and $32a_0$ with 4 repetitions (down). In both structures amorphous and crystalline regions have the same lengths (regularly spaced superlattices). In table-1, there is a summary of all studied structures, with their periodicities and their total lengths. Calculations have been done for superlattices of 4 periodicities: $16a_0$, $24a_0$, $32a_0$, $48a_0$. For each periodicity three different system sizes were considered according to the repetition of the superlattice's periodicity: 2, 4 and 6. We have to mention here that each amorphous region is a unique configuration (we did not repeat the same pattern to build the total structure). Figure-2 shows the atomic potential energies across the direction normal to the interfaces. The potentials calculated for each structure are similar; this proves the reproducibility and the quality of the created structures. The energies of the atoms within the crystalline or the amorphous regions are always the same. Furthermore the energy barriers of the atoms localized at the interfaces are almost the same for all of them. A first interesting remark is that the Si atoms that belong to the amorphous silicon regions close to the interfaces exhibit higher atomic energy than the inner atoms in amorphous regions. This might related to the enhancement of phonon scattering at the interfaces.

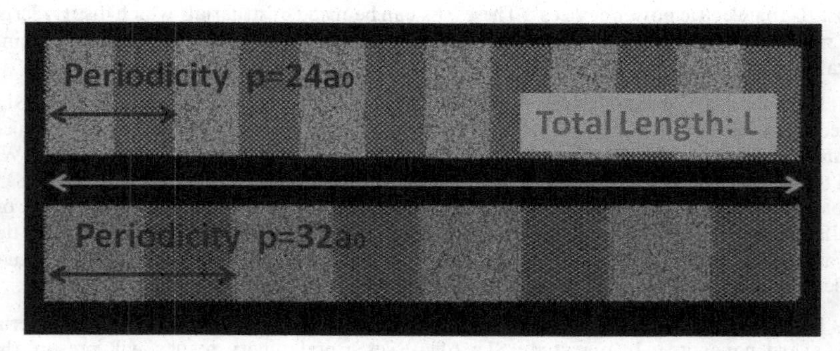


**FIGURE 1.** Superlattice structures: (up) period $24a_0$ or 48MLs and 6 repetitions, (down) $32a_0$ or 64MLs and 4 repetitions.

**Table I**. Periodicities and total lengths of studied superlattice's structures

| periodicity | Repetition | $16a_0$ | $24a_0$ | $32a_0$ | $48a_0$ |
|---|---|---|---|---|---|
| **Length Small (a0)** | 2 | 32 | 48 | 64 | 96 |
| **Length Medium (a0)** | 4 | 64 | 96 | 128 | 192 |
| **Length Large (a0)** | 6 | 96 | 144 | 192 | 288 |

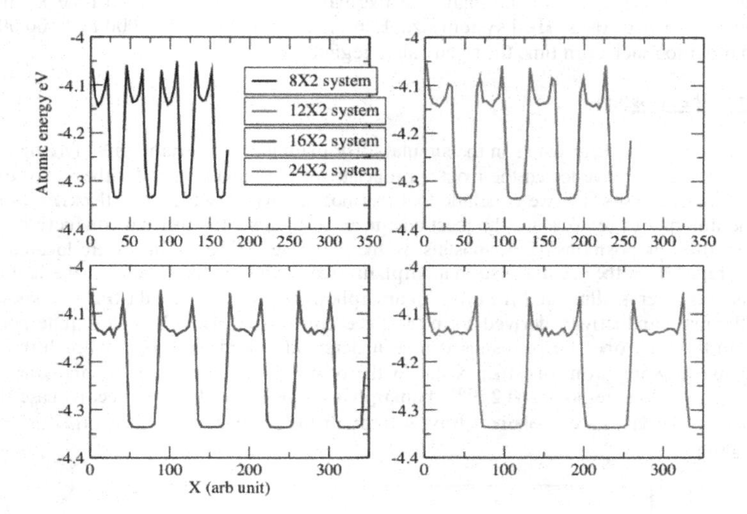


**FIGURE 2.** Atomic potential energies along the z direction, which is normal to the interfaces, of the four different studied amorphous/crystalline silicon superlattices.

The thermal conductivities of a series of regular portion of amorphous/crystalline silicon superlattices are obtained with a second molecular dynamics code, which is the well known simulation tool LAMMPS[7]. The Non-Equilibrium Molecular Dynamics method (NEMD) has been used to extract the heat flux and the temperature profiles. Then, the thermal conductivity is appraised on the basis of the Fourier's law[8,9] formalism. The thermal conductivity of an infinite system length is derived with the extrapolation method proposed by Shelling et al[10].

## DISCUSSION-RESULTS

Concerning the simulation procedure used to obtain the thermal conductivity, first simulations were done to define the effect of the thermostats size and of the time step on the appraised thermal conductivity. For all simulations achieved, we use thermostat sizes of ten mono-layers (MLs) or equivalent to the thickness of 10 MLs in the cases of amorphous regions. Time step of 4 fs are used in all simulation. For each structure we first run a NPH ensemble for 200.000 time steps and then a NVE to obtain the thermal conductivity. The total time for the NVE procedure depends of the studied system size. Here, it varies between 500.000 to 2.500.000 time steps, in order to reach each time the steady state regime.

### Location of the thermostats

The location of the thermostats in the simulation domain is an important issue. During the calibration procedure, we have tested the most appropriate location and keep it for the rest of our simulations. With this first study we conclude that thermostats made by the same thicknesses of crystalline and amorphous regions are the most adequate. In figure-3 the thermal conductivity is given for three different positions of thermostats. In the first case the thermostats were located in the crystalline regions, in the second case in amorphous ones, while in the third case the half of the thermostat was in crystalline and the other in amorphous region. The red dashed line shows the average thermal conductivity derived from the three considered cases. It is in a quite good agreement with the third one. These results can be understood considering the total path that a phonon feels, when going from one thermostat to the other. In the first case the phonons go through 1/3$^{rd}$ of crystalline regions and 2/3$^{rds}$ of amorphous regions, while in the second case the inverse occurs. In the last case a phonon travels through the same portion at amorphous and crystalline regions.

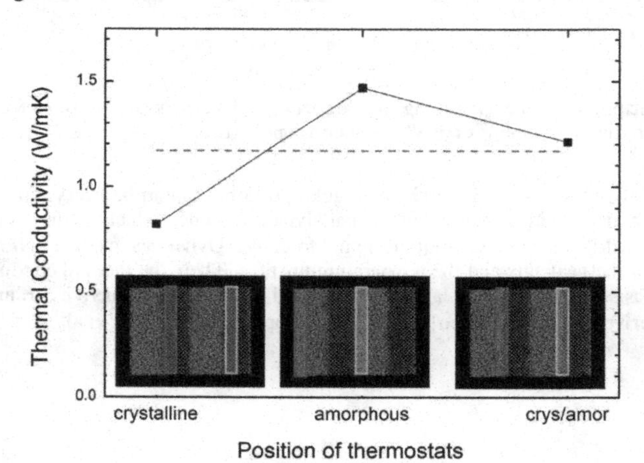


**FIGURE 3.** Thermal conductivity for structures with 2 repetitions of superlattice's periodicity equal to $16a_0$ at 300K versus the thermostats location.

## Thermal conductivity function the system size and the periodicity

In figure-4 the thermal conductivity is given as a function of the superlattice's periodicity and of the number of repetitions, while in figure-5 the thermal conductivity for an infinite structure length versus the periodicity is depicted. When the periodicity and the number of repetitions are increased, the thermal conductivity increases. Here we want to draw the attention on the effect of the number of interfaces on the thermal conductivity. Let's compare a first sample with a total length of 128 MLs containing 4 repetitions of a superlattice with periodicity of 32MLs with a second sample with the same total length but build by 2 repetitions of a superlattice periodicity of 64MLs. We see that the thermal conductivity changes even if the total lengths of crystalline and amorphous regions in each case are the same. Here, two phenomena contribute to this variation: when passing from the 2 repetitions to 4, the number of interfaces doubles and the existence of a finite Kapitza resistance leads to a decrease of the global conductivity. On the other hand, because the length traveled by phonons between two interfaces decreases when the periodicity changes from 64MLs to 32MLs, the effective conductivity of the superlattice layers decreases. Oppositely, these two effects tend to increase the conductivity when the periodicity is larger (64MLs). The simulations results conclude that this rise is of the order of 15%. As we will see later on, in the section dealing with Kapitza resistance, the $R_K$ value is 10% of the total thermal resistance. On the basis of this simple example we show that the thermal boundary resistance of the "virtual" interfaces between crystalline and amorphous regions is playing an important role for the overall thermal conductivity.

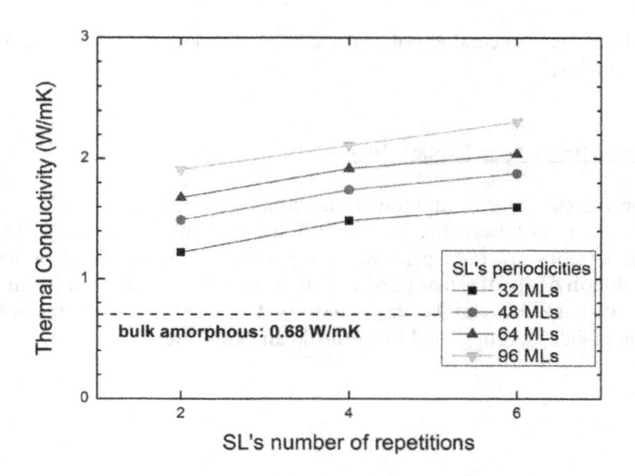


**FIGURE 4.** Thermal conductivity for the 4 SL's periods and 2, 4, and 6 periods is given at 300K.

Eventually, in figure-5, we show that there is a good agreement between the thermal conductivity obtained by NEMD and Equilibrium Molecular Dynamics (EMD) techniques. The fact that the thermal conductivity is not strongly affected by the periodicity tends to prove that the amorphous regions control the thermal conductivity of modeled the structures. The phonon mean free path is confined by the length of the half periodicity for each structure, which corresponds to the length of one crystalline region.

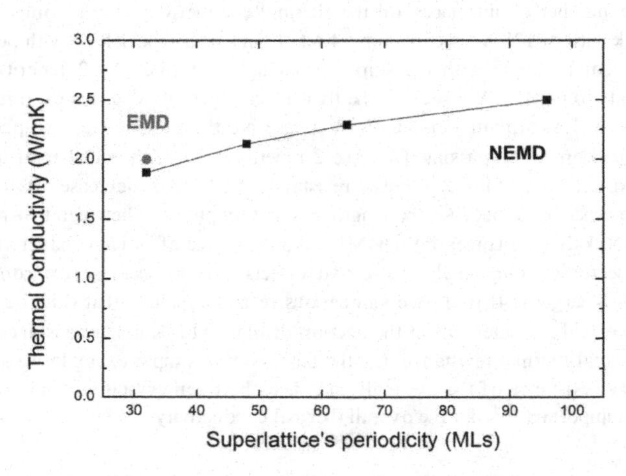


**FIGURE 5.** Extrapolated thermal conductivity for the SL's with four periodicities and 2, 4, and 6 repetitions is given at 300K.

**Thermal conductivity function the temperature**

In this section the structure mean temperature and especially how it affects the thermal conductivity of superlattices is investigate. In figure-6, the thermal conductivity of the smallest studied system (2 repetitions of $16a_0$ periodicity superlattices) is given as function of the temperature. The evolution of the thermal conductivity is weak and follows a linear trend. This unusual temperature dependence can be explained by the presence of a majority of non-propagating vibrational modes as suggested by Donadio and Galli [ref5].

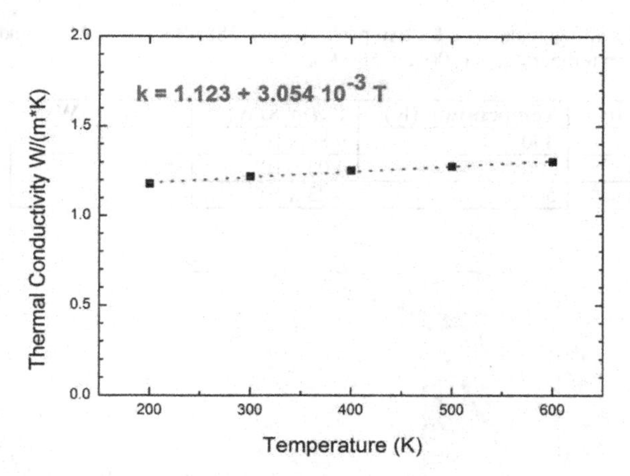


**FIGURE 6.** Thermal conductivity versus the temperature.

**Kapitza resistance function the temperature**

At nanoscales interfacial aspects can be dominant. In the case of multilayered materials it has been shown that the Kapitza resistance considerably affects their thermal conductivity. Measuring the temperature drop at the interface and knowing the heat flux through the samples we can calculate the Kapitza resistance $R_K$ using the Fourier's law. This Kapitza resistance was appraised for two cases, first for the smallest studied system (figure-7, 32MLs periodicity) and in the case of a large structure (periodicity 288 MLs). In both cases a single interface is considered. For the small structure at mean temperature of 300K, the $R_K$ value is equal to $4.03 \times 10^{-10}$ $m^2K/W$, while the total thermal resistance is $4.27 \times 10^{-9}$ $m^2K/W$. At 500K, the $R_K$ is equal to $4.22 \times 10^{-10}$ $m^2K/W$, while the total thermal resistance does not change. Thus, $R_K$ slightly increases with temperature; however this effect is weak and needs to be confirmed with supplementary calculations. In the case of the large system at 300K, the $R_K$ value is equal to $9.86 \times 10^{-10}$ $m^2K/W$. Here there is obviously a size effect and a dependence of the thermal boundary resistance with the considered number of monolayers. However, we currently cannot explain this phenomenon that also needs further investigations. Besides, we have to point out that well-converged results were obtained for the case of small periodicity structure, while for the large one the results are preliminary and more work is needed to be sure of their reliability. Eventually, the temperature independence of the total thermal resistance is consistent with the thermal conductivity given in figure-6 and discussed previously.

**Table II**. Caclulated Kapitza resistance for two periodicities (288MLs and 32MLs) and for the case of 32MLs, for two temperatures (300 and 500 K).

| periodicity | Temperature (K) | $R_K(m^2K/W)$ | $R_{Tot}(m^2K/W)$ |
|---|---|---|---|
| **288 MLs** | 300 | $9.86 \times 10^{-10}$ | $8.27 \times 10^{-9}$ |
| **32 MLs** | 300 | $4.03 \times 10^{-10}$ | $4.27 \times 10^{-9}$ |
| **32 MLs** | 500 | $4.23 \times 10^{-10}$ | $4.27 \times 10^{-9}$ |

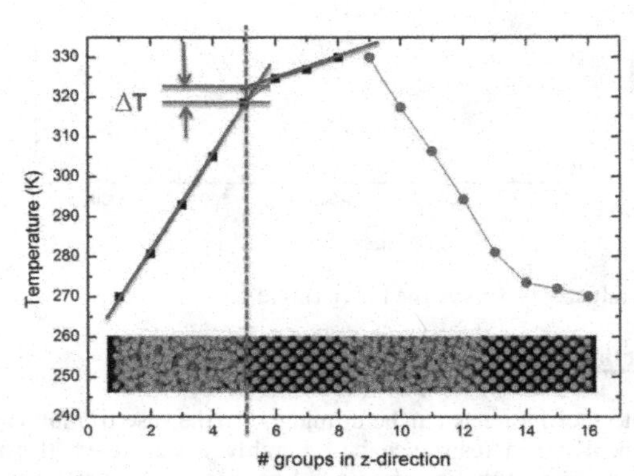


**FIGURE 7.** Temperature profile and the temperature jump at crystalline/amorphous interfaces.

## CONCLUSIONS

The thermal conductivity of amorphous/crystalline silicon superlattices was studied with the NEMD method. First, it shall be noticed that the quality of crystalline/amorphous interfaces is a major issue to consider, in order achieving reliable simulations. The thermal conductivity of the amorphous bulk structure obtained with the same methodology is 0.7 W/mK, while for crystalline bulk we found the value of 138W/mK at 300K. Comparing these values with the cross-plane thermal conductivity of the regular amorphous/crystalline superlattices, we see that the latter ones are close to the amorphous bulk silicon. This is an interesting result, if we consider the huge anisotropy that one can obtain between in-plane and cross-plane thermal conductivity with such systems. This property can be very useful in the case of thermoelectric devices design. Increasing the periodicity of regular a/c-Si-SLs has a weak effect on the cross-plane thermal conductivity. In contrast the in-plane thermal conductivity will considerably vary when the supelattice's periodicity is changed. The temperature dependence of the thermal conductivity of such systems shows also that the majority of modes are non-propagating. The Kapitza resistance of such structure interfaces was also appraised. The obtain results are coherent but further investigations are needed to propose a detailed interpretation of phonon interactions in such a/c-Si-SLs.

Now we plan the study of superlattices with different portions of crystalline and amorphous areas. Even if it has been proved that the interfaces between a crystalline and an amorphous material is very flat, we will try to introduce the effect of roughness of small thickness. Finally, the phonon density of states and the transmission coefficients will elucidate the physics hided in such interfaces.

## REFERENCES

1. R.W. Fathauer, "New class of Si-based superlattices: Alternating layers of crystalline Si and porous amorphous $Si_{1-x}Ge_x$ alloys", *Appl. Phys. Lett.* 61, 2350 (1992).
2. H. Ohra, R. Huang and Y. Ikuhara, "Large enhancement of the thermoelectric Seebeck coefficient for amorphous oxide semiconductor superlattices with extremely thin conductive layers", *Phys. Status Solidi (RRL)* **2**, 105(2008).
3. S.C. Agarwal, "Amorphous silicon-based superlattices ", *Bull. Mater. Sci.* **14**, 1257 (1991).
4. S. Von Alfthan, A. Juronen, and K. Kaski, *Mat. Res. Soc. Symp. Proc.* **703**, 6.2.1 (2002).
5. D. Donadio and G. Galli, "Temperature dependence of the thermal conductivity of thin silicon nanowires" *Nano Lett.* **10**, 847 (2010)
6. C. Fusco, T. Albaret and A. Tanguy, *Physical Review E* **82**, 066116 (2010)
7. S. J. Plimpton, R. Pollock, and M. Stevens, in Proceedings of the Eighth SIAM Conference on Parallel Processing for Scientific Computing (SIAM, Minneapolis, Minnesota, 1997).
8. K. Termentzidis, P. Chantrenne, and P. Keblinski, Phys. Rev. B **79**, 214307 (2009).
9. K. Termentzidis, S. Merabia, P. Chantrenne, and P. Keblinski, Int. J. Heat Mass Transf. **54**, 2014 (2011).
10. P. K. Schelling, S. R. Phillpot, and P. Keblinski, Phys. Rev. B **65**,144306 (2002).

# Nanostructured Bulk and Composites

Mater. Res. Soc. Symp. Proc. Vol. 1543 © 2013 Materials Research Society
DOI: 10.1557/opl.2013.972

# Thermoelectric properties of Ru and In substituted misfit-layered $Ca_3Co_4O_9$

Gesine Saucke[1], Sascha Populoh[1], Nina Vogel-Schäuble[1], Leyre Sagarna[1], Kailash Mogare[2], Lassi Karvonen[1] and Anke Weidenkaff[1,2]
[1]Empa - Swiss Federal Laboratories for Materials Science and Technology, Laboratory for Solid State Chemistry and Catalysis, Überlandstr. 129, CH-8600 Dübendorf, Switzerland
[2]University of Berne, Department of Chemistry and Biochemistry, CH-3012 Berne, Switzerland

## ABSTRACT

As an approach to improve the thermoelectric properties of the polycrystalline $Ca_3Co_4O_9$ misfit-layered oxide, substitutions of $Co^{2+\cdots4+}$ with the heavier cations $Ru^{3+/4+}$ and $In^{3+}$ were tested. Polycrystalline samples $Ca_3Co_{4-x}Ru_xO_9$ and $Ca_3Co_{4-x}In_xO_9$ ($0 < x < 0.21$) were prepared via a solid-state-reaction route. For each sample the crystal structure was analyzed and a complete thermoelectric characterization was done within a temperature range of 300 K < $T$ < 1125 K.

Both substitution strategies resulted in a significant decrease of the thermal conductivity ($\kappa$). For the $In^{3+}$-substituted samples the decrease of the Seebeck coefficient ($\alpha$) balanced the κ reduction so that no overall enhancement of the figure of merit ($ZT$) was found. The $Ru^{3+/4+}$ substitution reduced the p-type carrier concentration and thus increases the electrical resistivity ($\rho_{el}$), while α became larger at low temperatures. Despite the reduction of the power factor, a small enhancement in $ZT$ was observed in the case of $x = 0.1$ Ru substitution, due to the strong κ reduction. Considering the observed preferred orientation of the Ru-substituted crystallites, a maximum value of $ZT = 0.14$ perpendicular to the pressing direction is found at $T = 1125$ K, indicating that Ru substitution is a promising strategy for a further $ZT$ improvement.

## INTRODUCTION

The increasing demand for alternative and sustainable energies and energy conversion technologies leads to a growing interest in the field of thermoelectricity. Yet, thermoelectric converters are commercially available only for niche applications at low temperatures ($T$), since the expensive material or production costs prevent their extensive application for waste-heat recovery. Furthermore, the conversion efficiency increases with the applied temperature difference and thus an application at high temperatures is desirable. Compared with the present-day commercial modules, oxide materials have a high potentiality for waste-heat recovery at elevated temperatures. They possess the necessary chemical and high-temperature stability, promise low production and material costs and are non-toxic [1, 2].

One of the most promising thermoelectric p-type oxides is the misfit-layered $Ca_3Co_4O_9$ [3], characterized by an outstandingly high Seebeck coefficient ($\alpha$) at elevated temperatures [4-6]. The misfit-layered structure, combining electrically conductive [$CoO_2$] layers intergrown with insulating $Ca_2CoO_3$, provides a very low thermal conductivity ($\kappa$) and an acceptable electrical resistivity ($\rho_{el}$). A successful approach for a further improvement of the thermoelectric figure of merit ($ZT$) is cationic substitution [5, 7, 8]. Substitutions may improve α or $\rho_{el}$ through modifying the charge-carrier concentration, the band structure or the electron correlation effects [7]. Furthermore, the substitution with heavier cations may decrease $\kappa$ through introducing mass fluctuations and defects acting as phonon scattering centers in the crystal lattice [7, 5]. In the present work, the heavy cations $In^{3+}$ and $Ru^{3+/4+}$ are introduced into the lattice in order to decrease $\kappa$.

## EXPERIMENTAL PROCEDURE

Polycrystalline $Ca_3Co_4O_9$, $Ca_3Co_{4-x}Ru_xO_9$ ($x$ = 0; 0.1; 0.2) and $Co_3Co_{4-x}In_xO_9$ ($x$ = 0; 0.1; 0.21) powders were synthesized via conventional solid-state-reaction route (SSR) starting from $CaCO_3$, $Co_3O_4$, $In_2O_3$ or $RuO_2$ powders. Stoichiometric amounts of reactants were mixed and ball-milled (400 rpm for 1 h) using iso-octane as dispersing medium, whereupon the originally intended composition of $Ca_{3-x}In_xCo_4O_9$ ($x$ = 0; 0.1; 0.2) was considered. Pelletized (Perkin Elmer hydraulic press, pellet diameter 13 mm, pressure 70 kN) samples were placed in alumina boats and calcined in air at 900 °C for 24 h. The calcined pellets were ground manually using agate mortar and pestle, re-pelletized under similar conditions and sintered in air at 920 °C for 24 h (pristine and In-substituted samples) or 48 h (Ru-substituted samples). The as-sintered pellets were annealed in oxygen flow for 36 h with one intermittent grinding. X-ray measurements were done using a powder diffractometer (XRD; PANalytical X'pert PRO MRD; Cu-K$\alpha_1$ radiation, $\lambda$ = 0.15406 nm). The recorded diffraction patterns were fitted using LeBail method as implemented in the program FullProf.

For investigating the thermoelectric properties, each pellet was cut into two slices: one part was used for the determination of the thermal diffusivity ($\lambda$), while the second part was cut into bars for the $\alpha$ and the $\sigma$ measurements. Pieces left over from cutting were used for the specific heat ($C_p$) measurements. All temperature dependent measurements were performed in $O_2$ flow. $\alpha$ and $\sigma$ were measured using the Ozawa Science RZ2001i system. The error limits $\Delta\sigma$ = 7 % and $\Delta\alpha$ = 5 % for the system were estimated based on the errors in the measurement of the distance between the electrode-contact wires and the sample dimensions [9]. For the determination of $\kappa$ ($= \rho \lambda C_p$), $\lambda$ was measured using Laser Flash Absorption (LFA; Netzsch LFA 457 Microfash) and $C_p$ using a Differential Scanning Calorimeter (DSC; Netzsch DSC 404 C). The densities ($\rho$) of the samples were determined through weighing and measuring the geometric dimensions. The error $\Delta\lambda$ was calculated based on the measurement uncertainty and the inhomogeneity in the thickness of the samples (2-5 %). For the $C_p$ three independent measurements suggest a standard deviation $\Delta C_p$ < 5 % over the whole temperature range. The error $\Delta\rho$ was additionally considered as 2-8 % depending on the sample. The error of the figure of merit $\Delta ZT$ = 13-19 % follows from error propagation.

Hall measurements were performed in the Physical Property Measurement System (PPMS; Quantum Design) using the AC Transport option with a frequency of 1 Hz. The magnetic field strength was varied between -25 and +25 kOe, while the excitation currents were between 10 and 300 mA. For the Hall measurements the statistical error of the Hall-resistivity-magnetic-flux fitting and the errors in the measurement of the dimensions (8 %) were considered.

Finally, the morphology and the mean particle size of the different samples were investigated with scanning electron microscopy (SEM; Hitachi S-4800 and XL30 ESEM (FEI)) with a secondary electron detector. The SEM images were recorded at a breaking edge with a perspective perpendicular to the pressing direction.

## RESULTS AND DISCUSSION

The XRD-patterns (Fig. 1 a)) evidence the phase formation of the layered cobalt oxides, and all recorded patterns were LeBail fitted (FullProf) using the space group *C 2/m* with two independent b axis parameters of the misfit-layered structure. In addition to the oxide a minor impurity phase of $Co_3O_4$ $(Fd\bar{3}m)$ was found and two weak reflexes (marked by arrows) were unidentified. All of the LeBail fits had a good statistical quality ($1.57 < \chi < 3.05$, $1.01 < R_p < 1.28$, $1.35 < R_{wp} < 1.89$, $1.04 < R_{exp} < 1.09$). Nevertheless, the asymmetric peak shape especially of the c-axis reflexes leads to a nonzero asymmetric difference plot.

As presented in Tab. 1 and the inset of Fig. 1 a) for both substitutions, a shift of the lattice parameters is observed which suggests an inclusion of the substitution atoms into the lattice. In the case of the $Ru^{3+/4+}$ substitution the c axis parameter is expanded and no impurity phase was observed, which confirms a substitution of $Co^{2+...4+}$ [10] with $Ru^{3+/4+}$. In contrast, for an increasing $In^{3+}$ substitution the amount of the $Co_3O_4$ phase increases. Also the c-axis is expanded although for the substitution of the big Ca atom with the smaller In the opposite would be expected. Both of these findings support the idea that In substitutes the Co site instead of the attempted Ca site, which leads to the observed Co excess. As no In-based impurities were observed, the initial precursor mixture should lead to the composition $Ca_3Co_{4-x}In_xO_9$ ($x = 0$; 0.1; 0.21).

**Table I:** Relative density $\rho_{rel}$ in comparison to the theoretical density, as determined by geometrical measurements and weighing of the sample pellets, and lattice parameter determined by Lebail fitting (FullProf).

| | $\rho_{rel}$ (%) | $a$ (Å) | $b_1$ (Å) | $b_2$ (Å) | $c$ (Å) | $\beta$ (°) |
|---|---|---|---|---|---|---|
| $Ca_3Co_4O_9$ | 75(1) | 4.832(1) | 4.572(1) | 2.808(1) | 10.834(1) | 98.11(2) |
| $Ru_{0.1}$ | 72(1) | 4.834(1) | 4.569(1) | 2.807(1) | 10.851(1) | 98.10(2) |
| $Ru_{0.2}$ | 69(5) | 4.837(1) | 4.563(1) | 2.812(1) | 10.868(1) | 98.10(2) |
| $In_{0.1}$ | 71(1) | 4.828(1) | 4.563(1) | 2.812(1) | 10.873(1) | 98.11(2) |
| $In_{0.2}$ | 71(1) | 4.821(1) | 4.555(1) | 2.817(1) | 10.880(1) | 98.13(2) |

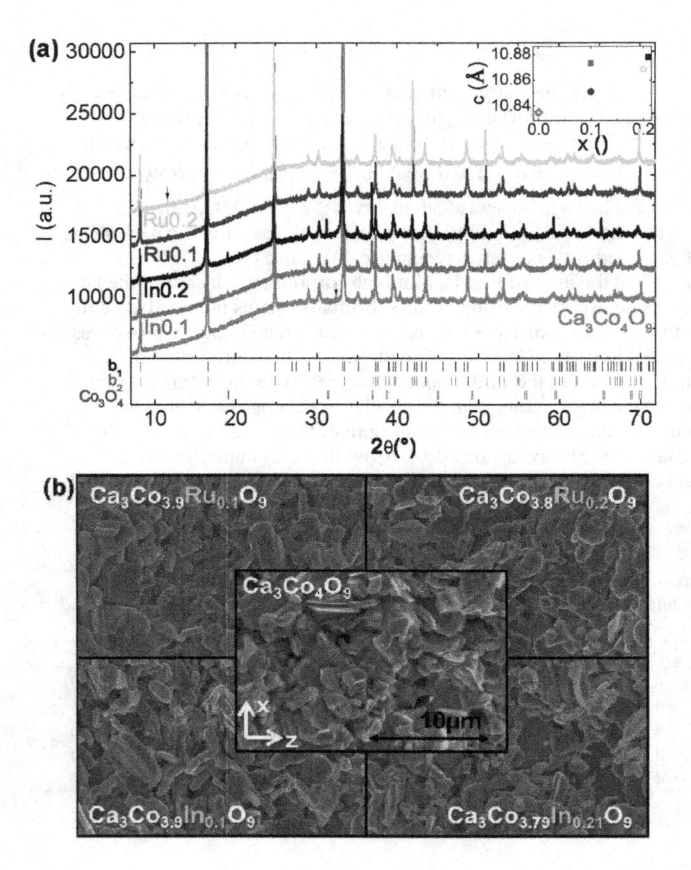


**Figure 1:** (a) X-ray diffraction pattern for different In and Ru substitution levels. Peak positions presented in the bottom were calculated via LeBail fitting (FullProf) of the $Ca_3Co_4O_9$ sample in the space group *C 2/m* with two independent b axis parameters and $Co_3O_4$ with the symmetry $Fd\overline{3}m$. Two weak reflexes are unidentified and marked by arrows. The Inset shows the determined expansion of the c lattice parameter due to the substitution ($Ru^{3+/4+}$: circles, $In^{3+}$: squares). Samples are named according to the substitution atom and the substitution level in the whole manuscript. (b) SEM micrographs of $Ca_3Co_4O_9$-based sintered pellets with different substitution levels showing the micro structure of the samples with a perspective perpendicular to the pressing direction z.

The micro structure recorded via scanning electron microscopy (SEM) is presented in Fig. 1 b). The images show very similar features for all samples: plate-like crystallites with thicknesses between 0.2-0.9 μm and diameters in the range of 0.4-4 μm. XRD-patterns of the pellet with $x = 0.1$ Ru-substitution were recorded to identify a possible preferred orientation of the crystallites with respect to the uniaxial pressing direction z,. First, the pellet was measured from its plain top side, and then it was cut into lamella which were turned by 90° in order to measure in the perpendicular direction (inset in Fig. 2 a)). The comparison of the XRD patterns recorded in the perpendicular directions (Fig. 2 a)) reveals a preferred orientation of the c-axis along z, which is obvious from the alteration in the intensity of the [00l] reflection intensities in comparison to the intensities of the reflections related to the other crystal planes. SEM micrographs taken in the two directions show different characteristics due to the alignment of the grains (Fig. 2 b)). While a plain view perspective parallel to the pressing direction (polished surface) reveals regular shaped particles with diameters between 0.5-4 μm, the plate like character of the particles with thicknesses in the range of 0.3-0.9 μm is only recognizable from the cross sectional view.

Weighing and geometrical measurements indicated similar densities between 3.25 and 3.45g/cm$^3$ for all samples. Using the measured lattice parameters and a substitution dependent unit cell mass, the relative densities ($\rho_{rel}$, relative to theoretical densities) are presented in Tab. I.

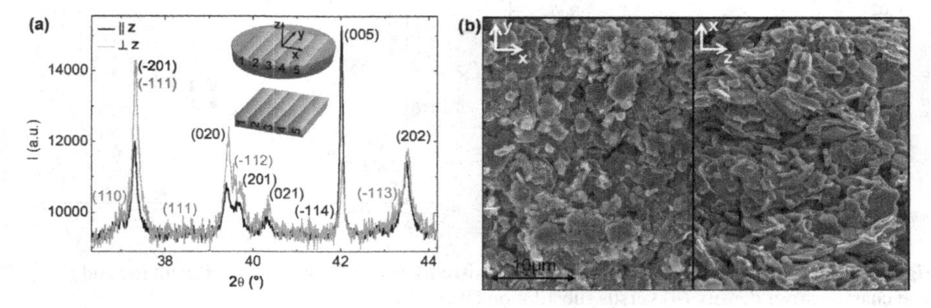


**Figure 2:** Preferred orientation of the crystallites in the pellet due to the uniaxial pressing direction z during fabrication. (a) Intensity change of XRD intensities for a measurement parallel and perpendicular to z. For measurements perpendicular to z the pellet was cut into bars, which were turned by 90° (inset, defined coordinate system valid for whole manuscript). (b) Alignment of crystallites perpendicular to pressing direction z obvious from a comparison of SEM images recorded in perpendicular directions ($x = 0.1$ Ru–substituted sample).

For all samples the electrical resistivity $\rho_{el}$ presented in Fig. 3 a) shows metal like behavior at low temperatures in agreement with the typical electrical transport behavior along the [$CoO_2$] layers [10]. Additionally, a transition between the metal like behavior at low temperatures and an activated transport above $T \approx 340$ K for the $x = 0.0$ , $T \approx 385$ K for the $x = 0.1$ In-substitution, $T \approx 527$ K for the $x = 0.21$ In-substitution and $T \approx 430$ K for both Ru substitutions is observed. This transition may be interpreted in terms of a spin-state transition, as discussed for the unsubstituted compound in Ref. [10].

With an increasing substitution of $Co^{2+...4+}$ with $Ru^{3+/4+}$ $\rho_{el}$ is increased over the whole temperature range, while α is increased only at $T < 650$ K (Fig. 3 b)). As $\rho_{el} \sim n^{-1}$ ($n$ = charge-carrier density) and $\alpha \sim n^{-2/3}$[11], both of these trends indicate a reduction of holes for $T < 650$ K, which is confirmed by room temperature Hall measurements shown in the Inset of Fig. 3 b). However, at high temperatures the $Ru^{3+/4+}$ doping decreases the Seebeck coefficient. The implied increase of the number of charge carriers can either be provided by a substitution of $Co^{2+}$ with $Ru^{3+}$ or $Ru^{4+}$ in the rock-salt layer or by a substitution of $Co^{3+/4+}$ with $Ru^{4+}$ in the conducting $[CoO_2]$ layer. The latter one can additionally reduce the mobility of the charge carriers in the $[CoO_2]$ layer through impurity scattering, which could further contribute to the strong increase of $\rho_{el}$.

For the substitution of $Co^{2+...4+}$ with $In^{3+}$ no significant change of $\rho_{el}$ is observed. Room temperature Hall measurements reveal a slight increase of n with increasing In substitution which is in agreement with the decrease of $\rho_{el}$ and the decrease of $\alpha$ (Fig. 3 b)) at this temperature. However, these trends are within the error range.

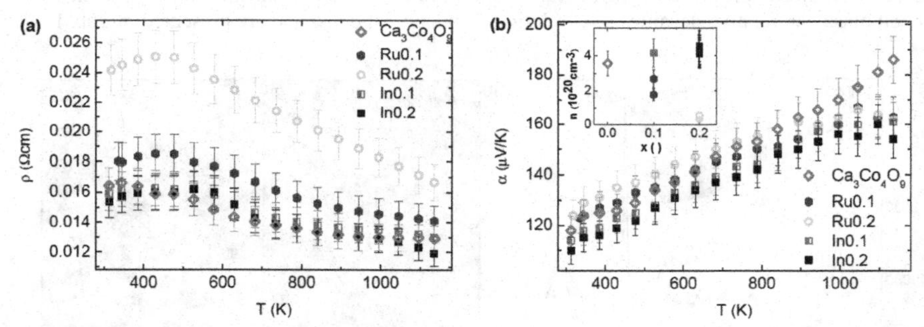


**Figure 3:** (a) Temperature dependent electrical resistivity ($\rho_{el}$), (b) Seebeck coeffcient ($\alpha$) and *p*-type charge carrier density ($n$) versus substitution level ($x$) (inset).

The thermal properties of the samples are presented in Fig. 4. In Fig. 4 a) the temperature dependence of the $C_p$ is shown. Theoretically, the isobaric molar specific heat for solids

$$c_p \approx c_v = \underbrace{\frac{12\pi^4 R}{5}\left(\frac{T}{T_D}\right)^3}_{c_{V,ph}} + \underbrace{\frac{\pi^2 R}{2}\frac{T}{T_F}}_{c_{V,el}} + c_{v,an} \tag{1}$$

is approx. equal to the isochoric heat $c_V$ which is determined by the sum of a phononic $c_{V,ph}$, an electronic $c_{V,el}$ and an anharmonic contribution $c_{V,an}$, whereupon R is the gas constant for an ideal gas, $T_D$ the Debye temperature and $T_F$ the Fermi temperature. Above $T_D$ the phononic contribution can be approximated by $c_{V,ph} = 3R$ according to the law of Dulong-Petit. This contribution to the $C_p = c_p/M_{av}$ is marked as horizontal lines in Fig. 4 a), whereas $M_{av}$ is the average atomic mass of the compound. In accordance with the Debye theory for all samples the $C_p$ increases with $T$ and reaches or exceeds the Dulong-Petit limit around 600-700 K, which is in good agreement with an observed Debye temperature of $T_D$ = 660 K found for $Ca_3Co_4O_9$ [12]. At higher temperatures a further increase of the $C_p$ values is observable, which could be explained by electronic or anharmonic contributions. A comparison of the different degrees of substitution reveals that the substitution does not have a significant influence on $C_p$ for In and $x = 0.1$ Ru substitution. However, for the sample with $x = 0.2$ Ru substitution the increase of $C_p$ at high temperatures seems to be less pronounced in accordance with a lower charge carrier concentration.

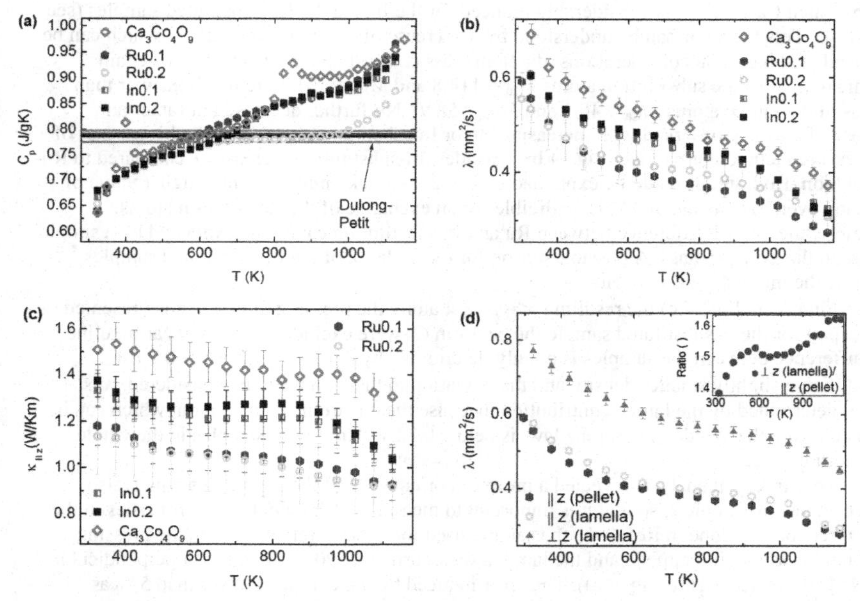


**Figure 4:** (a) Specific heat capacity values ($C_p$) as measured with DSC (symbols) and calculated by Dulong-Petit (lines). (b) Thermal diffusivities ($\lambda$). (c) Thermal conductivities parallel to z ($\kappa_{\|z}$). (d) Thermal diffusivity of the $x = 0.1$ Ru substituted sample parallel and perpendicular to the pressing direction z.

The thermal diffusivity $\lambda = 1/3v_g^2\,\tau$, which depends on the phonon group velocity $v_g$ and the relaxation time $\tau$, decreases with increasing temperature of all samples (Fig. 4 b)). This general temperature trend is in agreement with Umklapp-phonon scattering for which the relaxation time

$$\tau_U(\omega) \propto \frac{M_{av} v_g v_p^2}{\gamma^2 T} \tag{2}$$

is determined by $v_g$, the phase velocity $v_p$ of the phonons, the Grüneisen parameter $\gamma$ and the average atomic mass $M_{av}$ of the compound and decreases with $T$ [13]. In comparison to the unsubstituted $Ca_3Co_4O_9$ $\lambda$ is considerably reduced for the In and the Ru substituted samples (see Fig. 4 b)). This behavior can be understood by a decrease of the group velocity $v_g$, which can be predicted [13] based on considerations of a high mass contrast along a 1-dim atomic chain (atomic masses of the substitution atoms $M_{In} = 114.8$ and $M_{Ru} = 101.1$ are much heavier than masses of the lattice atoms $M_{Ca} = 40.1$ and $M_{Co} = 58.9$). No further decrease but rather an increase of $\lambda$ can be observed with increasing Ru or In substitution level over the bigger part of the measured temperature range (Fig. 4 b)). Besides, In-substitution increases $\lambda$ compared to Ru-substitution. Both trends could be explained by Eq. 2 if we take into account a higher value of $M_{av}$. However, the change of $M_{av}$ is negligible for an exchange of the substitution atoms. Therefore, the large $\lambda$ difference between Ru and In substitution cannot be explained by a simple change in the average mass. A possible reason for the higher $\lambda$ of the In-substituted samples might be the minor $Co_3O_4$ impurity.

Finally, in Fig. 4 c) the resulting $\kappa = \rho\, C_p\, \lambda$ along the pressing direction z is presented, whereupon for the unsubstituted sample the peaks in $C_p$ were excluded using a linear baseline. The difference between the samples is mainly determined by $\lambda$. However, for $x = 0.2$ Ru substitution a slightly smaller density and the deviation in the $C_p$ have to be considered. $\kappa$ is mainly determined by the lattice contribution, because the electronic contribution, which can be estimated from the Wiedemann-Franz law, is below 0.22 W/Km for all samples in the whole temperature range.

The structural analysis revealed a preferred orientation of the crystallites which also reflects in an anisotropic $\lambda$, so that it is important to measure all thermoelectric properties in the same direction. As done in Ref. [14], the sample used for $\lambda$ measurements along the pressing direction z was cut into lamella and the lamella were turned by 90 ° to measure $\lambda$ perpendicular to z [15] (compare Inset of Fig. 2 a)). The error induced by the cutting is lower than 5 % as apparent from the comparison of the diffusivity measurement of the uncut sample with the cut lamella (both along the pressing direction) (Fig. 4 d)). In the inset, the ratio $(\lambda_{||z})/(\lambda_{\perp z})$ of the pellet diffusivity (parallel to the pressing direction) to the perpendicular measured diffusivity (lamella turned by 90 °) is presented. This temperature dependent ratio was used to calculate $\kappa$ along the x-y direction for all samples. In Fig. 4 d) a comparison of $\lambda$ measured for the whole pellet along the pressing direction z and perpendicular to z are presented, revealing a higher $\lambda$ in the perpendicular direction. This observation is in agreement with the alignment of the plate-like crystallites, because on the one hand the relaxation time increases, due to less grain-boundary scattering of the phonons in the x-y plane compared to z, and on the other hand because of a higher $\lambda$ within the crystallites parallel to the layers. A similar behavior was found in Ref. [16].

Finally, in Fig. 5 the figure of merit *ZT*, calculated from transport measurements along the x-y direction, is presented for the different substitutions. While for the In substitution no enhancement of *ZT* was observed, $x = 0.1$ Ru substitution leads to a slight increase of *ZT*, due to the strong reduction of $\kappa$. At 1125 K a maximum $ZT = 0.14$ was reached for the $x = 0.1$ Ru substituted sample. Consequently, continuing investigations with a lower Ru substitution levels is a promising strategy for a further improvement of *ZT*.

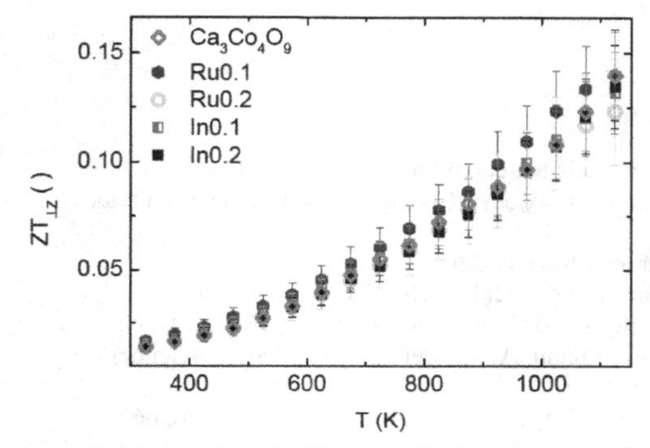


**Figure 5:** Figure of merit *ZT* perpendicular to the pressing direction. $\lambda$ was measured perpendicular to the pressing direction for the $x = 0.1$ Ru substituted sample and calculated for the other samples based on the $(\lambda_{||z})/(\lambda_{\perp z})$ factor determined in Fig. 4 b).

## CONCLUSIONS

Polycrystalline $Ca_3Co_{4-x}Ru_xO_9$ and $Ca_3Co_{4-x}In_xO_9$ ($0.0 < x < 0.21$) were synthesized through solid-state-reaction route. Structural analysis revealed a preferred orientation of the crystallites due to the uniaxial pressing of the samples, which led to an anisotropic $\lambda$. In accordance with Umklapp-phonon scattering, being the dominant scattering mechanism, $\kappa$ decreased for In and Ru substitution. An increasing Ru-substitution decreased the carrier density, which led to a decrease of $\sigma$ and an increase of $\alpha$ at low temperatures. In contrast, In-substitution did not show a clear influence on $\sigma$, while $\alpha$ is reduced and the carrier density seemed to be slightly increased. All in all, for $x = 0.1$ Ru substitution an increase of *ZT* was observed due to the strong decrease of $\kappa$. A maximum value of $ZT = 0.14$ at $T = 1125$ K was obtained. Considering the slight increase of *ZT*, Ru substitution seems to be a promising strategy for further improvements.

## ACKNOWLEDGMENTS

Funding was achieved by the Bundesamt für Energie (BFE) under Grant No. Si/500601 and vonRoll casting (emmenbrücke) ag, Rüeggisingerstrasse 2, CH-6020 Emmenbrücke). Furthermore, the authors want to thank Wenjie Xie, Song Hak Yoon and Krzysztof Galazka for fruitful discussions.

## REFERENCES

1. J. He, Y. Liu, R. Funahashi, Journal of Materials Research 26(15), 1762 (2011).
2. D.W. Bruce, D. O'Hare, R.I. Walton, Functional oxides - Chapter 4: Thermoelectric Oxides (Wiley, Chichester, 2010)
3. J.W. Fergus, Journal of the European Ceramic Society 32(3), 525 (2012).
4. M. Shikano, R. Funahashi, Applied Physics Letters 82(12), 1851 (2003).
5. N. Nong, C. Liu, M. Ohtaki, Journal of Alloys and Compounds 491(1-2), 53 (2010).
6. D. Moser, L. Karvonen, S. Populoh, M. Trottmann, A. Weidenkaff, Solid State Sciences 13(12), 2160 (2011).
7. Y. Wang, L. Xu, Y. Sui, X. Wang, J. Cheng, W. Su, Applied Physics Letters 97(6), 062114 (2010).
8. Q. Yao, D.L. Wang, L.D. Chen, X. Shi, M. Zhou, Journal of Applied Physics 97(10), 103905 (2005).
9. S. Populoh, M. Aguirre, O. Brunko, K. Galazka, Y. Lu, A. Weidenka_, Scripta Materialia 66(12), 1073 (2012).
10. A.C. Masset, C. Michel, A. Maignan, M. Hervieu, O. Toulemonde, F. Studer, B. Raveau, J. Hejtmanek, Physical Review B 62(1), 166 (2000).
11. G.J. Snyder, E.S. Toberer, Nature Materials 7(2), 105 (2008).
12. Y. Wang, Y. Sui, J. Cheng, X. Wang, W. Su, Journal of Physics: Condensed Matter 19(35), 356216 (2007).
13. E.S. Toberer, A. Zevalkink, G.J. Snyder, Journal of Materials Chemistry 21(40), 15843 (2011).
14. W. Xie, J. He, S. Zhu, T. Holgate, S. Wang, X. Tang, Q. Zhang, T.M. Tritt, Journal of Materials Research 26(15), 1791 (2011).
15. For the laser flash measurement the lamella were put into a square-shaped sample holder and the upper side was coated with an Ethanol - $Ca_3Co_4O_9$ slurry, sprayed on the lamella with an airbrush (A709 by AZTEK), to fill the small gaps between the lamella. The cobalt oxide powder for the slurry was produced through a soft chemistry method. .
16. L. Zhang, N. Okinaka, T. Tosho, T. Akiyama, Journal of Optoelectronics and advanced materials 14(1-2), 67 (2012)

**Mater. Res. Soc. Symp. Proc. Vol. 1543 © 2013 Materials Research Society**
**DOI: 10.1557/opl.2013.948**

# Enhanced Thermoelectric Figure-of-merit at Room Temperature in Bulk Bi(Sb)Te(Se) With Grain Size of ~100nm

Tsung-ta E. Chan[1], Rama Venkatasubramanian[1], James M. LeBeau[2], Peter Thomas[1], Judy Stuart[1] and Carl C. Koch[2]

[1]Center for Solid State Energetics, RTI International, Research Triangle Park, NC 27709, U.S.A.

[2]Department of Materials Science and Engineering, North Carolina State University, Raleigh, NC 27606, U.S.A.

## ABSTRACT

Grain boundaries are known to be able to impede phonon transport in the material. In the thermoelectric application, this phenomenon could help improve the figure-of-merit (ZT) and enhance the thermal to electrical conversion. $Bi_2Te_3$ based alloys are renowned for their high ZT around room temperature but still need improvements, in both n- and p-type materials, for the resulting power generation devices to be more competitive. To implement high density of grain boundaries into the bulk materials, a bottom-up approach is employed in this work: consolidations of nanocrystalline powders into bulk disks. Nanocrystalline powders are developed by high energy cryogenic mechanical alloying that produces Bi(Sb)Te(Se) alloy powders with grain size in the range of 7 to 14 nm, which is about 25% finer compared to room temperature mechanical alloying. High density of grain boundaries are preserved from the powders to the bulk materials through optimized high pressure hot pressing. The consolidated bulk materials have been characterized by X-ray diffraction and transmission electron microscope for their composition and microstructure. They mainly consist of grains in the scale of 100 nm with some distributions of finer grains in both types of materials. Preliminary transport property measurements show that the thermal conductivity is significantly reduced at and around room temperature: about 0.65 W/m-K for the n-type BiTe(Se) and 0.85 W/m-K for the p-type Bi(Sb)Te, which are attributed to increased phonon scattering provided by the nanostructure and therefore enhanced ZT about 1.35 for the n-type and 1.21 for the p-type are observed. Detailed transport properties, such as the electrical resistivity, Seebeck coefficient and power factor as well as the resulting ZT as a function of temperature will be described.

## INTRODUCTION

With the advantages such as reliability, solid state operation (no moving parts) and zero emission, thermoelectric (TE) cooling or power generation devices have emerged in the search of "green" energy sources. The main challenge of the TE devices is to improve the efficiency of the conversion between heat and electricity, which is mainly determined by the thermoelectric materials' figure of merit (ZT), defined as $ZT=\alpha^2T/\rho k$, where $\alpha$, T, $\rho$, k are the Seebeck coefficient, absolute temperature, electrical resistivity and thermal conductivity, respectively. Therefore ZT can be enhanced by increasing the Seebeck coefficient and/or decreasing the electrical resistivity and thermal conductivity.

To reduce thermal conductivity, the concept of utilizing nanostructure to scatter phonons has been proposed and experimentally confirmed in thin film PbTe, BiTe materials and other material geometries [1,2]. In bulk materials, similar concept was adapted and resulted in a "nanocomposite" structure, i.e., nanoscale particles/grains embedded in the matrix materials. Nanocomposite BiSbTe and SiGeP materials with increased grain boundary density was found effective in the lattice thermal conductivity reduction [3–6]. To further improve ZT, it would be desirable to have as many nanometer scale grains as possible. Cryogenic milling, an alternative method of the more common room temperature ball milling, could provide another route to achieve such nanocrystalline structure. This technique is able to produce nano-scale microstructure in various metals and alloys much more efficiently than conventional room temperature mechanical alloying [7–12]. The reduced grain size in the resulting powders are expected to somewhat compensate the grain growth during the consolidation process and maintain a high density of grain boundaries to scatter phonons and reduce the thermal conductivity.

## EXPERIMENT

The bulk materials were obtained by consolidating mechanical alloyed powders using uniaxial hot press. Elemental powders, Bi, Sb, Se and Te, supplied by Alfa Aesar of purity 99.99% or higher of appropriate weight ratio were loaded into the stainless steel vials with stainless steel balls, ball to powder ratio was 10:1, inside a glove box with oxygen less than 1 ppm. The vials were then vacuumed and sealed. The mechanical alloying was carried out by a high energy cryogenic milling system: the motor provides 1725 RPM and clamp speed 1060 cycles/minute. Before the milling, the vial was bathed in liquid nitrogen for 20 minutes and received a continuous flow of liquid nitrogen during the milling time of 4 hours. The resulting powders were then consolidated at temperatures from 400 to 450 °C with various pressures from 0.8 to 2.5 GPa in an inert gas environment. The resulting bulk materials are typically 800 $\mu$m in thickness and 10 mm in diameter.

The x-ray diffraction (XRD) spectra were obtained for both powder and bulk samples by a Rigaku GeigerFlex DMax/A diffractometer with a Cu radiation. The density of bulk samples was measured by the Archimedes method. A JEOL 2000 transmission electron microscope (TEM) operated at 200 kV was used to acquire microstructure images and selected area electron diffraction patterns (SAD). The TEM specimens from powder samples were made by particle solution suspension while those from bulk samples were prepared by mechanical polishing followed by low energy ion milling.

The electrical resistivity of the materials was measured by the well-known van der Pauw method [13] in a Hall-effect set up. The Seebeck coefficients were measured based on the slope of the voltage versus temperature difference curves, as discussed in round-robin measurements [14]. The thermal conductivity of nano-bulk samples were measured in the same direction as the electrical resistivity and the Seebeck coefficient, with measured heat flow ($Q$) using carefully calibrated $Q$-meter, and then calculated from the Fourier law with measured temperature difference, cross section area and length of the sample.

## DISCUSSION

### Compositions

Mechanical alloyed powders at cryogenic temperature as well as hot-pressed bulk materials were analyzed by X-ray diffraction. The X-ray spectra in Figure 1 indicate the compositions of the materials well matched the targeted n-type BiTe(Se) and p-type Bi(Sb)Te and there is no second phase present in the powders or bulk compactions. It was found that n-type $Bi_2Te_{2.7}Se_{0.3}$ and p-type $Bi_{0.4}Sb_{1.6}Te_3$ deliver the best outcomes in transport properties.

The broadened peaks in the as-milled powders suggest the majority of the grains in the structures are in nanometer scale, which are then confirmed by the TEM images. After the hot pressing, the grains grow so that the XRD peaks are sharper and more peaks can be identified.

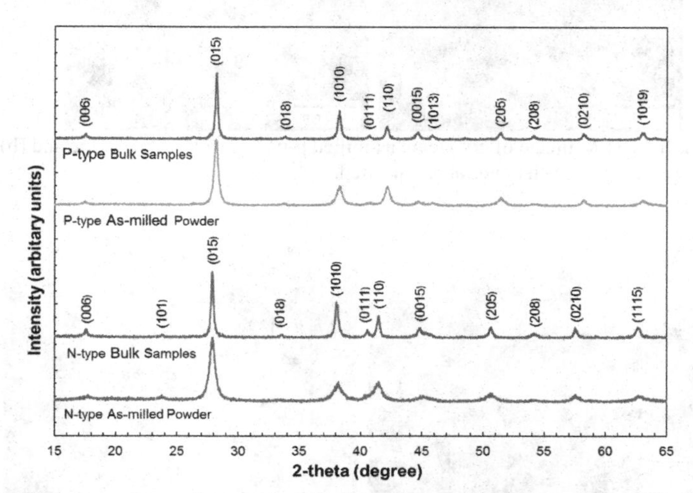


Figure 1. XRD spectra of n- and p-type as-milled powders and bulk compactions. The intensity is in arbitrary units.

## Microstructures

The dark field TEM images of as-milled powders are shown in Figure 2. The majority of the grains have size around 16 ± 5 nm for the p-type and 9 ± 3 nm for the n-type nanocrystalline powders. The nanostructure is also confirmed by the SAD pattern inserted in Figure 2. The spots almost form complete rings indicating small grains in the structure. The TEM images in Figure 3 of as-pressed bulk p- and n-type materials reveal that the nanocrystalline structures in the as-milled powders are partially maintained in the bulk materials. Both p-and n-type materials mainly consist of grains smaller than 100 nm surrounded by large grains about 200 nm or larger. The grains are closely packed, which are consistent with full density measurement results.

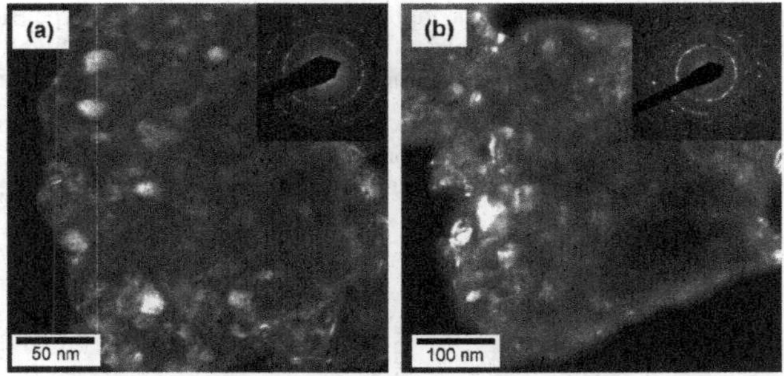

Figure 2. Dark field TEM image of cryogenic as-milled p-type powders. (a) n-type and (b) p-type. The SAD patterns of each material are inserted.

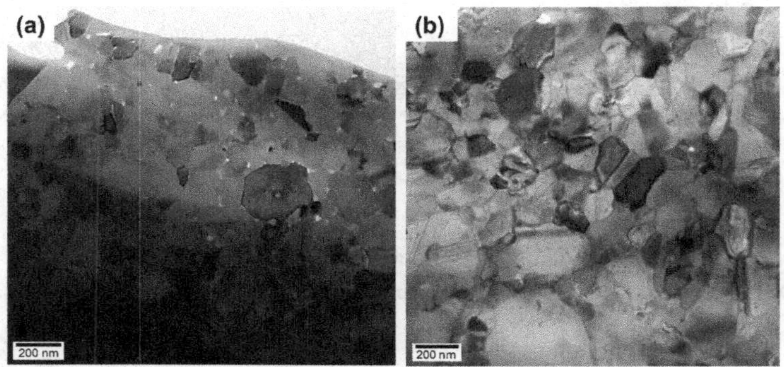

Figure 3. TEM images of bulk as-pressed materials: (a) n-type and (b) p-type materials.

**Transport properties**

Enhanced ZT values around room temperature have been observed: about 1.35 for the n-type and 1.21 for the p-type materials. Their individual properties are shown in Figure 4. It is noticed that the lattice thermal conductivity is reduced, mostly in the lattice contribution part as shown in Figure 4(f). The lattice thermal conductivity was calculated according to the Wiedemann-Franz law, $k_l = k_T - L\sigma T$, where $k_l$, $k_T$, $\sigma$, $T$ and $L$ are the lattice thermal conductivity, total thermal conductivity, electrical conductivity, absolute temperature and Lorentz number ($2.25 \times 10^{-8}$ $W\Omega K^{-2}$ in the case of a semiconductor). This phenomenon was attributed to the increased interface density in the microstructure, e.g. grain boundary density, which impedes phonon transport. On the other hand, increased interface density also results in much higher electrical resistivity starting with $1.7 \times 10^{-5}$ Ω-m at 25 °C and increases as temperature increases. This was compensated by the slightly improved Seebeck coefficient and reduced thermal conductivity. The improvement in the Seebeck coefficient could be from quantum confinement effect in the small grains as well as energy filtering effect. When the scale of the structure is down to nanometer scale, low energy charge carriers could be scattered by the interfaces so that the average energy of the charge carriers increases and thus increases the Seebeck coefficient, in addition to the quantum confinement effect that changes the electronic density of state distribution of the charge carriers [15,16].

On the other hand, it is noticed that the transport properties of both n- and p-type materials show similar temperature dependency which indicates a good matching factor for future device development.

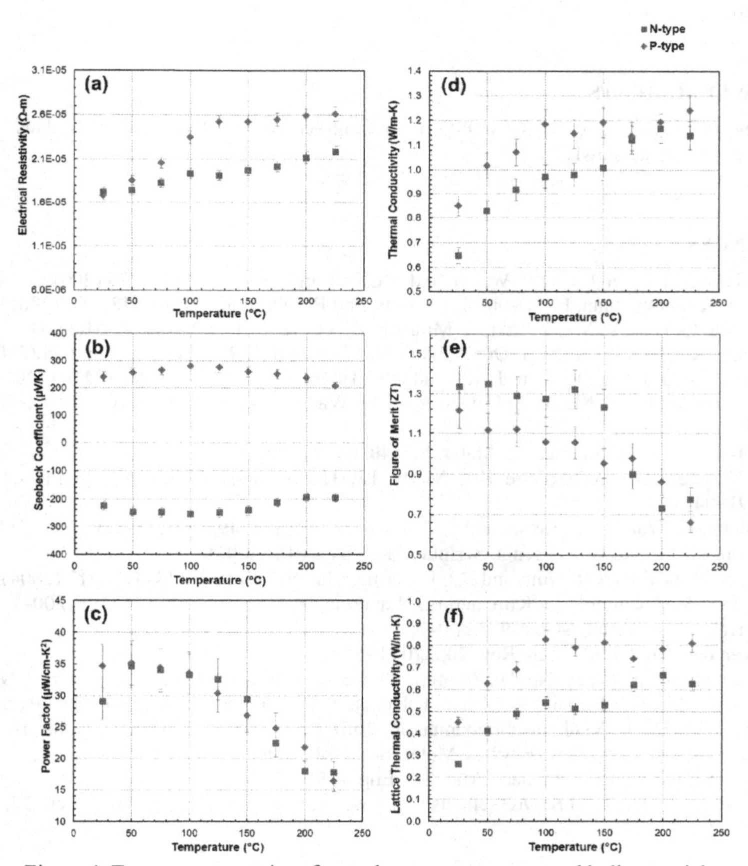


Figure 4. Transport properties of n- and p-type nanostructured bulk materials.

## CONCLUSIONS

Bulk $Bi_2Te_3$ based alloys with enhanced thermoelectric figure of merit around room temperature was developed. By introducing a high density of nanstructure, the lattice thermal conductivity was effectively reduced by interface scattering of phonons. Both TEM and X-ray diffraction have shown such nanocrystalline structure. The increased interface density may also be responsible for increased electrical resistivity and Seebeck coefficient. Therefore there is still room to optimize the power factor, $\alpha^2/\rho$, as well as the total thermal conductivity by modifying the nanostructure. On the other hand, device development with nanostructured p- and n-type materials, given the similarity in their properties as a function of temperature, is under investigation.

## ACKNOWLEDGMENTS

Research is funded by DARPA/DSO Army Contract No. W911NF-08-C-0058. This support is gratefully acknowledged.

## REFERENCES

1. T.C. Harman, P.J. Taylor, M.P. Walsh, and B.E. LaForge, Science **297**, 2229 (2002).
2 R. Venkatasubramanian, E. Siivola, T. Colpitts, and B. O'Quinn, Nature **413**, 597 (2001).
3. B. Poudel, Q. Hao, Y. Ma, Y. Lan, A. Minnich, B. Yu, X. Yan, D. Wang, A. Muto, D. Vashaee, X. Chen, J. Liu, M.S. Dresselhaus, G. Chen, and Z. Ren, Science **320**, 634 (2008).
4. S. Fan, J. Zhao, J. Guo, Q. Yan, J. Ma, and H.H. Hng, Appl. Phys. Lett. **96**, 182104 (2010).
5. X. Yan, B. Poudel, Y. Ma, W.S. Liu, G. Joshi, H. Wang, Y. Lan, D. Wang, G. Chen, and Z.F. Ren, Nano Lett. **10**, 3373 (2010).
6. O. Yamashita and S. Sugihara, J. Mater. Sci. **40**, 6439 (2005).
7. K.M. Youssef, R.O. Scattergood, K.L. Murty, J.A. Horton, and C.C. Koch, Appl. Phys. Lett. **87**, 091904 (2005).
8. X. Zhang, H. Wang, J. Narayan, and C.C. Koch, Acta Mater. **49**, 1319 (2001).
9. F. Zhou, J. Lee, and E.. Lavernia, Scripta Mater. **44**, 2013 (2001).
10. F. Zhou, D. Witkin, S.R. Nutt, and E.J. Lavernia, Mat. Sci. Eng. A **375–377**, 917 (2004).
11. Y.S. Park, K.H. Chung, N.J. Kim, and E.J. Lavernia, Mat. Sci. Eng. A **374**, 211 (2004).
12. C.C. Koch, Nanostruct. Mater. **9**, 13 (1997).
13. L.J. van der Pauw, Phil. Tech. Rev. **20**, 220 (1958).
14. N.D. Lowhorn, W. Wong-Ng, W. Zhang, Z.Q. Lu, M. Otani, E. Thomas, M. Green, T.N. Tran, N. Dilley, S. Ghamaty, N. Elsner, T. Hogan, A.D. Downey, Q. Jie, Q. Li, H. Obara, J. Sharp, C. Caylor, R. Venkatasubramanian, R. Willigan, J. Yang, J. Martin, G. Nolas, B. Edwards, and T. Tritt, Appl. Phys. A-Mater. Sci. Process. **94**, 231 (2009).
15. P. Pichanusakorn and P. Bandaru, Mat. Sci. Eng. R **67**, 19 (2010).
16. T. Koga, S.B. Cronin, M.S. Dresselhaus, J.L. Liu, and K.L. Wang, Appl. Phys. Lett. **77**, 1490 (2000).

Mater. Res. Soc. Symp. Proc. Vol. 1543 © 2013 Materials Research Society
DOI: 10.1557/opl.2013.932

# Impact of Rapid Thermal Annealing on Thermoelectric Properties of Bulk Nanostructured Zinc Oxide

Markus Engenhorst[1], Devendraprakash Gautam[1], Carolin Schilling[1], Markus Winterer[1], Gabi Schierning[1] and Roland Schmechel[1]
[1]Faculty of Engineering and Center for Nanointegration Duisburg-Essen (CENIDE), University of Duisburg-Essen, 47057 Duisburg, Germany

## ABSTRACT

In search for non-toxic thermoelectric materials that are stable in air at elevated temperatures, zinc oxide has been shown to be one of only few efficient *n*-type oxidic materials. Our *bottom-up* approach starts with very small (<10 nm) Al-doped ZnO nanoparticles prepared from organometallic precursors by chemical vapor synthesis using nominal doping concentrations of 2 at% and 8 at%. In order to obtain bulk nanostructured solids, the powders were compacted in a current-activated pressure-assisted densification process. Rapid thermal annealing was studied systematically as a means of further dopant activation. The thermoelectric properties are evaluated with regard to charge carrier concentration and mobility. A Jonker-type analysis reveals the potential of our approach to achieve high power factors. In the present study, power factors larger than $4\times10^{-4}$ Wm$^{-1}$K$^{-2}$ were measured at temperatures higher than 600 °C.

## INTRODUCTION

Oxides draw interest as candidate materials for thermoelectric energy conversion at elevated temperature in air because of their oxidation stability and the avoidance of elements, which are scarce, expensive or toxic. ZnO has been proposed as an *n-type* thermoelectric material almost 20 years ago [1] and today shows the highest conversion efficiencies known for *n-type* oxidic materials [2,3]. The dimensionless figure of merit

$$zT = \frac{\alpha^2 \cdot \sigma}{\kappa} \cdot T \quad (1)$$

is a characteristic measure for the conversion efficiency of a thermoelectric material. Herein $\alpha$ is the Seebeck coefficient, $\sigma$ the specific electrical conductivity, $\alpha^2\times\sigma$ the power factor, $\kappa$ the thermal conductivity, and $T$ the absolute temperature. Since the transport coefficients $\alpha$, $\sigma$ and $\kappa$ cannot be optimized independently, it is a common approach to lower the lattice thermal conductivity by means of nanostructuring [4,5], introducing grain boundaries as scattering centers for phonons, but not to the same extent for charge carriers.

One way of nanostructuring a material is a *bottom-up* process starting from nanoparticle powders that are compacted into nanocrystalline bulk material. Gas-phase synthesis is advantageous for the preparation of nanoparticles because their properties like diameter, size distribution and doping concentration can be tailored and the synthesis facilities could be upscaled to industrial levels [5,6].

A challenge during the compaction process is to preserve a high density of grain boundaries and to simultaneously obtain a high amount of electronically active dopant atoms. This is especially difficult in Al-doped ZnO (AZO) due to the solubility limit of the dopant which is estimated to be as low as between 0.3 and 0.5 at% [7,8], where 0.5 at% correspond to $1.8\times10^{26}$ dopant atoms per $m^3$ [9]. Excessive aluminum precipitates as the electrically inactive spinel phase $ZnAl_2O_4$ [9]. The present study investigates in how far rapid thermal annealing (RTA) with extremely fast heating and cooling rates (i) enables a higher dissolution of the dopant atoms into the host lattice, (ii) possibly freezes out additional dopant atoms kinetically and (iii) simultaneously preserves the nanostructure.

## EXPERIMENTAL METHODS

In a *bottom-up* approach, AZO nanoparticles were prepared by chemical vapor synthesis [6,10]. The liquid organometallic precursors diethylzinc and triethylaluminum were vaporized and carried into a hot wall reactor by helium gas. The bubbler temperatures and the ratio of the flow rates were adjusted to obtain nominal doping concentrations of either 2 at% or 8 at% Al in ZnO. Oxygen was used as reaction gas leading to the formation of the nanoparticles (< 10 nm) at a temperature of 900 °C and a pressure of 20 mbar. The powders were collected using thermophoresis.

Compaction was carried out in an *HP D 5/2* furnace (*FCT Systeme GmbH*). This furnace employs current-activated, pressure-assisted densification (CAPAD) enabling high heating and cooling rates. In this study, four pellets of 20 mm diameter were compacted at 35 MPa using graphite piston and punches. Heating and cooling rates were 100 K/min, dwell time was 3 minutes and the process chamber was purged with argon gas at a pressure of approximately 5 mbar. The dwell temperature was varied between 600 °C, 700 °C and 900 °C for three pellets with a nominal doping concentration of 8 at%. One pellet from 2 at% Al-doped powder was densified at 700 °C.

Rapid thermal annealing was carried out in a *MILA-5000-P-N* (*Ulvac-Riko, Inc.*) in a flow of 14 slm nitrogen at atmospheric pressure. Heating and cooling rates were 45 K/s, dwell time was 20 s. Dwell temperatures of 1000 °C and 1150 °C were consecutively applied on all samples.

Thermoelectric transport coefficients were determined between room temperature and 700 °C after both CAPAD and each annealing step. Seebeck coefficient and electrical conductivity were measured in a *ZEM-3* (*Ulvac-Riko, Inc.*) in Helium at 100 mbar. Thermal conductivity was determined by a laser flash method (*LFA 457 Microflash* from *NETZSCH-Gerätebau GmbH*) in a flow of 75 sccm $N_2$ at atmospheric pressure using a pyroceramic reference sample for calculation of the specific heat capacity. Density of the samples was measured exploiting the Archimedes principle.

## EVALUATION METHODS

Assuming a parabolic band and an energy-independent scattering approximation, the Seebeck coefficient of a degenerate semiconductor shows a linear dependence on the temperature as expressed in eq. (2) [4].

$$\alpha(T) = -\frac{8\cdot\pi\cdot k_B^2}{3\cdot e\cdot h^2}\cdot m^*\cdot T\cdot\left(\frac{\pi}{3\cdot n}\right)^{\frac{2}{3}} \qquad (2)$$

Here, $k_B$ denotes the Boltzmann constant, $e$ the elementary charge, $h$ the Planck's constant, $m^*$ the density-of-states (DOS) effective mass and $n$ the charge carrier concentration. For ZnO, $m^* = 0.27 \times m_e$ [11], with $m_e$ being the electron rest mass. Deriving $|\partial\alpha/\partial T|$ yields a constant slope from which the concentration of active charge carriers in the regarded temperature regime can be calculated [12].

Furthermore, from eq. (2) and the measured electrical conductivity

$$\sigma = n \cdot e \cdot \mu \tag{3}$$

the mobility $\mu$ can be estimated.

Another method to evaluate the doping efficiency was presented by Jonker [13] and applied to AZO by Ohtaki *et al.* [1]. According to these studies,

$$\alpha = \frac{k_B}{e}\ln\sigma - \frac{k_B}{e}\ln(N_c \cdot e \cdot \mu \cdot \exp A), \tag{4}$$

if hole conduction can be neglected. This means that the Seebeck coefficient linearly depends on the natural logarithm of the electrical conductivity as long as the transport mechanism (transport constant $A$), the effective conduction band DOS $N_c$ and the mobility are unaffected by changing the dopant level or the temperature. Consequently, $ln\ (N_c \times e \times \mu \times exp\ A)$ can be considered constant and expressed as $ln\ \sigma_0$. Plotting $\alpha$ versus $ln\ \sigma$ for samples which were subjected to the same process chain, i.e. CAPAD and RTA, but may contain different dopant concentrations, should generally allow for a determination of a common x-axis intercept $ln\ \sigma_0$ when fitting those samples with a linear regression of fixed slope $k_B/e$. Given a certain material it is beneficial to tailor its microstructural and electronic properties in a fashion that enlarges $ln\ \sigma_0$, which opens the possibility of achieving higher power factors, e.g. by increasing the mobility. The charge carrier concentration can be optimized subsequently to find the maximum power factor. Ohtaki *et al.* find a value of $ln\ \sigma_0 = 12.4$ for their AZO at 900 °C [1].

## RESULTS AND DISCUSSION

Figure 1 shows the thermoelectric properties for all samples measured at 150 °C after CAPAD, after RTA at 1000 °C and after RTA at 1150 °C. The charge carrier concentration and the mobility were estimated from eqs. (2) and (3).

The strong enhancement in electrical conductivity for the samples densified at 600 °C and 700 °C can be attributed to the considerable increase in both charge carrier concentration and mobility. Although the Seebeck coefficient decreased slightly due to the increase in charge carrier concentration, the power factor still shows a strong enhancement, except for the sample densified at 900 °C. This sample seems to lose active charge carriers by RTA, indicating that CAPAD might have activated more charge carriers in the material than are allowed in thermodynamic equilibrium.

It can be noted that all samples have very similar charge carrier concentrations between 1.4 and $1.8 \times 10^{26}$ $m^{-3}$ after RTA at 1150 °C confirming the low solubility limit of Al in ZnO cited in the introduction. Also, all samples exhibit similar mobility values between 30 and 38 $cm^2V^{-1}s^{-1}$ after RTA at 1150 °C. Interestingly, the latter mobility value is found for both samples densified at 700 °C no matter how much Al was added in excess. However, after RTA at 1000 °C the sample with a nominal Al-content of 2 at% shows a 1.8 times higher mobility than the sample with 8 at% Al-content. This might be attributed to a stronger suppression of grain growth caused by the spinel phase $ZnAl_2O_4$ [14] in the sample with higher Al-content.

Due to the nanocrystalline microstructure the thermal conductivity at 150 °C is very low for all samples. The higher the dwell temperature in CAPAD, the higher is the resulting thermal conductivity. By RTA the thermal conductivity of all samples increases. Comparing the increase in power factor and the increase in thermal conductivity due to RTA at 1000 °C, it can be noticed that the power factor is more enhanced than the thermal conductivity, leading to a higher figure of merit at 150 °C (not shown).

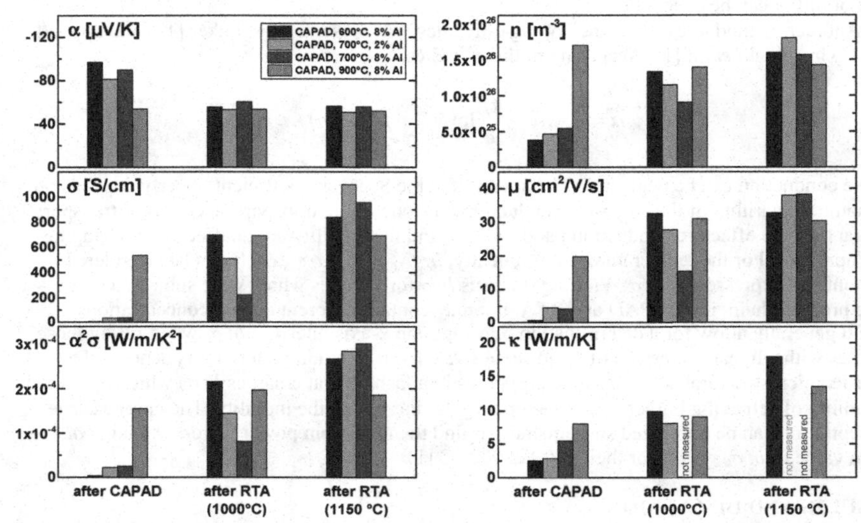


**Figure 1.** Thermoelectric transport properties for all samples measured at 150 °C after CAPAD, after RTA at 1000 °C and after RTA at 1150 °C. Charge carrier concentration and mobility were estimated from eqs. (2) and (3).

While this holds valid for lower temperatures, at higher temperatures the figure of merit slightly decreases due to the treatment by RTA. Exemplarily, the temperature dependence of the thermoelectric transport coefficients is shown in fig. 2 for all samples after RTA at 1000 °C. The largest differences between the samples can be seen in the electrical conductivity which can be attributed to the differences in charge carrier concentration and mobility as illustrated in fig. 1. After RTA at 1000 °C, the 2 % Al-doped sample densified at 700 °C exhibits a slightly lower thermal conductivity than the other two samples shown here. We suggest that grain growth, increase in grain connectivity and decrease in residual porosity depend on complex interdependent kinetics. Although there are differences in the transport coefficients, the overall figure of merit is quite similar for all samples illustrated in fig. 2, with the 2 at% Al-doped sample densified at 700 °C exhibiting a maximum $zT$ of 0.055 at 680 °C.

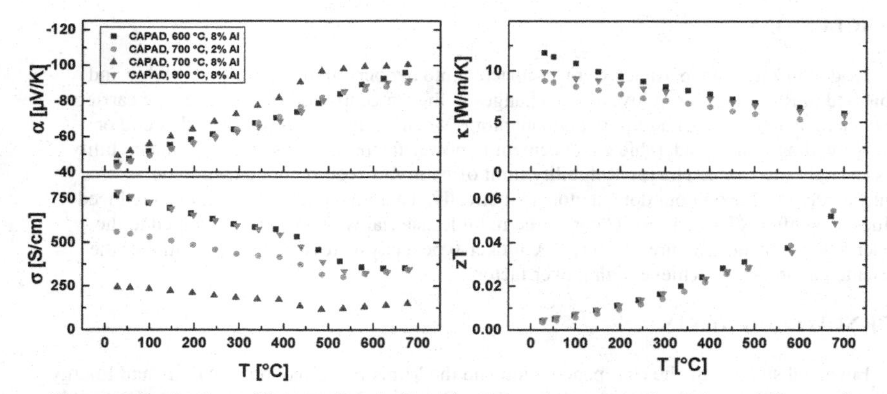


**Figure 2.** Seebeck coefficient, electrical conductivity, thermal conductivity and figure of merit for all samples after RTA at 1000 °C. Thermal conductivity of the 8 at% Al-doped sample densified at 700 °C was not measured.

In order to get a better insight how RTA influences the electrical transport properties with respect to the power factor, a Jonker-type analysis is performed on the two samples densified at 700 °C. Figure 3 reveals a shift of $ln\ \sigma_0$ towards higher values after each RTA step. After CAPAD, a common $ln\ \sigma_0 = 10.7$ can be attributed to both samples. A common x-intercept cannot be attributed to the samples after RTA at 1000 °C because they are significantly different at least with respect to mobility (see also fig. 2) which is one of the factors that is assumed constant when attributing a common $ln\ \sigma_0$. After RTA at 1150 °C, the x-intercept shifts to $ln\ \sigma_0 = 12.0$ which is close to the value of 12.4 found by Ohtaki *et al.* [1]. This demonstrates that the presented approach is capable of achieving high $ln\ \sigma_0$ which is necessary for good thermoelectric performance. The power factor corresponding to the values shown for both samples after RTA at 1150 °C amounts to $4.2\times10^{-4}$ $Wm^{-1}K^{-2}$. In order to achieve even higher power factors, it would be necessary to reduce the amount of active charge carriers which could possibly be done without affecting $ln\ \sigma_0$ too much.

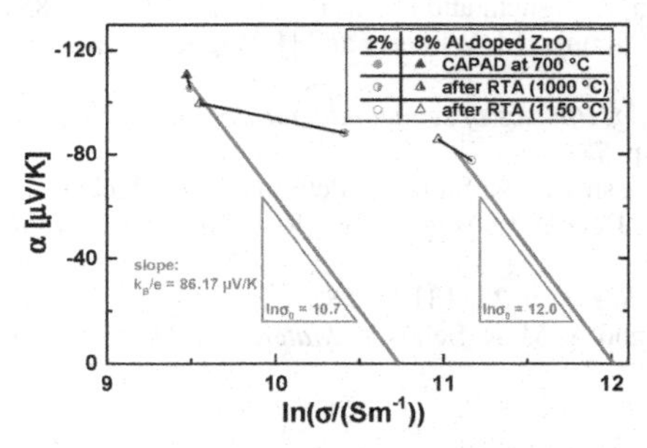


**Figure 3.** Jonker-plot for 2% and 8% Al-doped samples densified at 700 °C. The presented values were measured at 620 °C.

## CONCLUSIONS

Al-doped ZnO nanoparticles were compacted into nanocrystalline bulk by CAPAD and subjected to RTA in order to investigate changes in the concentration of active charge carriers. RTA yielded activation of additional dopant atoms when CAPAD was performed at 600 or 700 °C leading to a considerable enhancement in power factor. At the same time, the mobility was strongly increased. The low solubility limit of Al in ZnO could not be overcome significantly by freezing out dopant atoms kinetically. Thermal conductivity showed reduced values even after RTA at 1150 °C compared to bulk material which can be attributed to the preservation of nanostructured features. A Jonker-type analysis revealed the potential of the presented approach to achieve high power factors.

## ACKNOWLEDGMENTS

Financial support by the European Union and the Ministry of Economic Affairs and Energy of the State North Rhine-Westphalia in Germany (Objective 2 Programme: European Regional Development Fund, ERDF) is gratefully acknowledged. Financial support in the frame of a young investigator grant by the Ministry for innovation, science and research of the State North Rhine Westphalia in Germany is gratefully acknowledged.

## REFERENCES

1. M. Ohtaki, T. Tsubota, K. Eguchi and H. Arai, *J. Appl. Phys.* **79**, 1816 (1996).
2. M. Ohtaki, K. Araki and K. Yamamoto, *J. Electron. Mater.* **38**, 1234 (2009).
3. P. Jood, R. J. Mehta, Y. Zhang, G. Peleckis, X, Wang, R. W. Siegel, T. Borca-Tasciuc, S. X. Dou and G. Ramanath, *Nano Lett.* **11**, 4337 (2011).
4. G. J. Snyder and E.S. Toberer, *Nature Materials* **7**, 105 (2008).
5. K. Nielsch, J. Bachmann, J. Kimling and H. Böttner, *Adv. Energy Mater.* **1**, 713 (2011).
6. M. Ali, N. Friedenberger, M. Spasova and M. Winterer, *Chem. Vap. Dep.* **15**, 192 (2009).
7. K. Shirouzu, T. Ohkusa, M. Hotta, N. Enomoto and J. Hojo, *J. Ceram. Soc. Jpn.* **115**, 254 (2007).
8. D. Bérardan, C. Byl and N. Dragoe, *J. Am. Ceram. Soc.* **93**, 2352 (2010).
9. T. Tsubota, M. Ohtaki, K. Eguchi and H. Arai, *J. Mater. Chem.* **7**, 85 (1997).
10. S. Hartner, M. Ali, C. Schulz, M. Winterer and H. Wiggers, *Nanotechnology* **20**, 445701 (2009).
11. S. M. Sze and K. K. Ng, *Physics of semiconductor devices*, 3rd ed. (Wiley-Interscience, Hoboken, NJ, 2007), p. 789.
12. G. Schierning, R. Theissmann, N. Stein, N. Petermann, A. Becker, M. Engenhorst, V. Kessler, M. Geller, A. Beckel, H. Wiggers and R. Schmechel, *J. Appl. Phys.* **110**, 113515 (2011).
13. G. H. Jonker, *Philips Res. Rep.* **23**, 131 (1968).
14. J. Han, P. Q. Mantas and A. M. R. Senos, *J. Mater. Res.* **16**, 459 (2001).

Mater. Res. Soc. Symp. Proc. Vol. 1543 © 2013 Materials Research Society
DOI: 10.1557/opl.2013.947

# Fabrication and Characterization of Nanostructured Bulk Skutterudites

Mohsen Y. Tafti [1], Mohsin Saleemi[1], Alexandre Jacquot[2], Martin Jägle[2], Mamoun Muhammed[1], Muhammet S. Toprak[1]
[1]Department of Materials and Nano Physics, KTH - Royal Institute of Technology, Stockholm, Sweden
[2]Fraunhofer-Institute for Physical Measurements IPM, Freiburg, Germany

## ABSTRACT

Latest nanotechnology concepts applied in thermoelectric (TE) research have opened many new avenues to improve the ZT value. Low dimensional structures can improve the ZT value as compared to bulk materials by substantial reduction in the lattice thermal conductivity, $\kappa_L$. However, the materials were not feasible for the industrial scale production of macroscopic devices because of complicated and costly manufacturing processes involved. Bulk nanostructured (NS) TEs are normally fabricated using a bulk process rather than a nano-fabrication process, which has the important advantage of producing in large quantities and in a form that is compatible with commercially available TE devices.
We developed fabrication strategies for bulk nanostructured skutterudite materials based on $Fe_xCo_{1-x}Sb_3$. The process is based on precipitation of a precursor material with the desired metal atom composition, which is then exposed to thermochemical processing of calcination followed by reduction. The resultant material thus formed maintains nanostructured particles which are then compacted using Spark Plasma Sintering (SPS) by utilizing previously optimized process parameters. Microstructure, crystallinity, phase composition, thermal stability and temperature dependent transport property evaluation has been performed for compacted NS $Fe_xCo_{1-x}Sb_3$. Evaluation results are presented in detail, suggesting the feasibility of devised strategy for bulk quantities of doped TE nanopowder fabrication.

## INTRODUCTION

Thermoelectric (TE) materials have recently found a lot of interest and investment. They can directly interconvert between heat and electricity and have no moving parts, with almost close to zero noise [1-8]. In order to have good TE material, the material's thermal conductivity should behave similar to a glass and on the other hand have the electronic properties of a crystal [21]. This concept is called Phonon Glass Electron Crystal (PGEC). Skutterudites based on $CoSb_3$ are among the materials which may behave similarly to the PGEC concept. [9] The skutterudite structure is based on $MX_3$ where M is a transition metal (Fe, Co, Ni, etc.) and the X can be P, As, or Sb [1,4-6,8]. Skutterudite unit cell is represented by the face centered cubic arrangement of M atoms forming 8 sub-cubic voids/cages. Four membered rings of X atoms fill six of these eight cages. The empty two voids can allow introduction of heavy elements forming so called filled-skutterudites. Several strategies have been introduced to enhance the electronic properties as well as reducing the thermal conductivity of skutterudites including doping, filling, and nanoengineering. Doping can dramatically enhance the electrical conductivity as well as introducing point defects, which can reduce the thermal conductivity; on the other hand filling the voids in the crystal structure with rare earth elements can drastically reduce the thermal conductivity. Finally having nanostructured bulk material can additionally introduce a high density of grain boundaries and thus reduce the thermal conductivity further [1, 3-6, 8, 10-15].

There have been several synthesis strategies introduced for making these materials which the most commonly used routes that are metallurgical such as melt fusion, ball milling, chemical alloying and etc. Each of these routes provide us with benefits and unfortunately have some disadvantages; for example melt fusion can result in high purity material with different doping elements and fillings but the time consumed are in order of several days, and the amount of heat and energy used per batch are more than 700°C up to 1100°C -which are not desirable. Ball milling can provide with high impact energy which may assist in obtaining nanostructured materials and even assist in filling the rare earth elements into the cages, but doesn't result in the pure phase material due to impurities coming from the milling container and balls, and a second step of annealing is usually followed. Although structural filling of skutterudite with rare-earth atoms is not possible, chemical alloying on the other hand can provide narrowly distributed submicron particles, which result in high phonon scattering –thus a low thermal conductivity [1,3-7,9,13,15,17-20]. In this article we report on the use of chemical alloying route to produce iron doped skutterudite and present its structural, transport property (TE) evaluations as well as thermal cycling/stability.

## EXPERIMENT

### Thermodynamic modeling

In order to synthesize the precursor materials we have to first perform thermodynamic modeling of the synthesis process and chose the optimum parameters of synthesis according to the modeling. This will allow us to obtain the material with the right stoichiometry and highest reaction yield. For the thermodynamic modeling Medusa software has been used and the results are presented in Fig. 1 for the formation of oxalate salts of iron, cobalt and antimony. The working pH region for the precipitation of the precursor was obtained as 2.5 to 3.5 based on the modeling results.

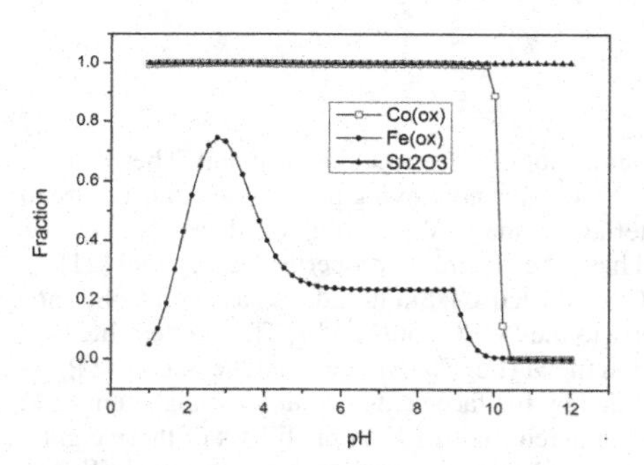


**Figure 1**. Thermodynamic modeling for precipitation of $Co^{2+}$, $Fe^{2+}$ and $Sb^{3+}$ species.

### Synthesis

For preparing 30 grams of final skutterudite powder, stock solutions of iron, cobalt and antimony were prepared by dissolving 48.43 g antimony chloride ($SbCl_3$), 5.92 g cobalt chloride ($CoCl_2$) and 1.69 g of iron chloride hydrate ($FeCl_2{\bullet}H_2O$) in 3M hydrochloric acid (HCl) solution (the powders were used as received from Sigma Aldrich) forming solutions of 0.75 M, 0.225 M, and 0.025 M solutions respectively. Taking the reaction yield of iron oxalate into consideration 20% extra iron chloride has been added to the calculations for compensation. Ammonium oxalate solution, to be used as the precipitating agent for all these three ions, was made with 0.3M. 3M HCl and 3M ammonium hydroxide solution were prepared for adjusting the pH at 3. The reaction was carried out at room temperature. The metal ion solution and the precipitating agent (ammonium oxalate) were added drop wise into the reaction chamber simultaneously until all the solutions are consumed and desired precipitates are formed. After the completion of the process the powders were filtered from the supernatant and washed with DI water several times, dried at 80°C over night. The dried powders were then processed in a box furnace (Carbolite Furnaces) at 350°C in air for about an hour to be followed by reduction under controlled reducing Hydrogen atmosphere at 450°C for 3 hours. The consolidation process was done by Spark Plasma Sintering (SPS Dr. Sinter 2050). Phase purity was determined by X-ray powder diffraction (Powder diffractometer Panalytical X'Pert PRO) and the morphology was evaluated with Scanning Electron Microscopy (SEM Zeiss Ultra 55).

## RESULTS AND DISCUSSION

Synthesis process has been successfully performed based on the modeling results at the estimated pH window. The dried powder is pinkish and the calcined powder is light grey, while the final powder is dark grey - allowing a visual follow-up of the process steps. The precursor material, calcined and reduced powders, as well as the consolidate pellets have been evaluated for their composition. Fig. 2 shows the XRD pattern of the final pellet. The pattern shows pure skutterudite phase meaning that all the iron elements have been doped in cobalt sites.

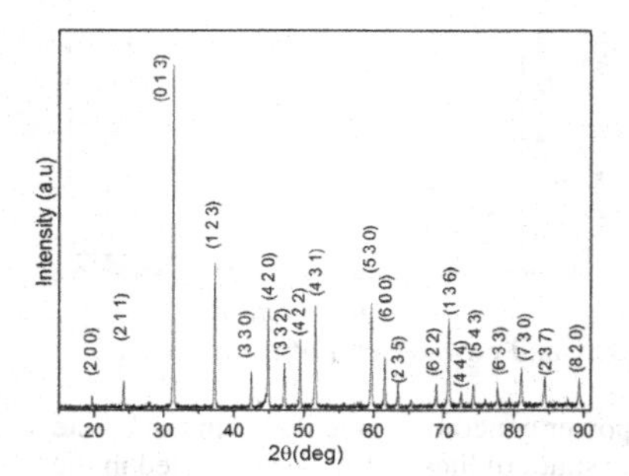


**Figure 2**. XRD pattern of SPS consolidated pellet indexed for $CoSb_3$ (ICCD# 01-083-0055)

With SPS parameters of 70°C/min ramp and 475°C as the sintering temperature we have achieved compaction densities of more than 96% of the theoretical density. Fig. 3 shows SEM

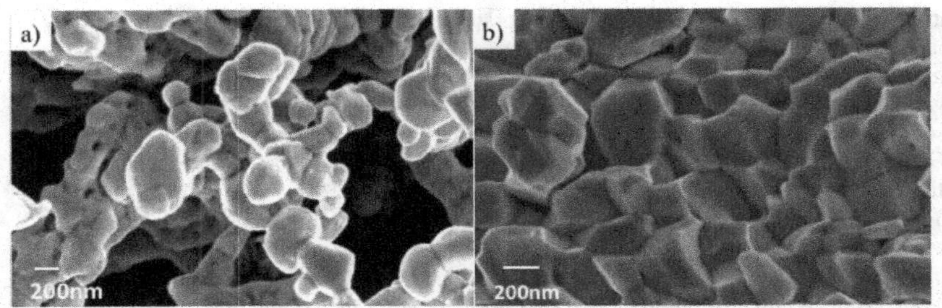


**Figure 3**. SEM micrograph of (a) as prepared powder, (b) SPS consolidated sample.

micrographs of the as prepared powder as well as the consolidated sample. Particle size of the reduced powder is in the range of 200 nm to 400 nm, while the consolidated sample exhibits grain sizes in the range of 250 nm to 600 nm with high compaction density. This indicated that there was a small impact of sintering conditions on the grain size of reduced powder.

Electrical conductivity and Seebeck measurements have been performed on sintered sample. The sample shows p-type behavior as reported earlier by Zhou et. al. [5] and Park et. al. [19]. It starts from 25 μV/K at room temperature increasing to 110 μV/K at 800 K. The electrical conductivity is about 1500 S/cm at room temperature and decrease by increasing temperature, reaching 700 S/cm at 800 K. This is about 1.5 times higher comparing to 600 S/cm for [19] and 650 S/cm [5]. The Seebeck coefficient is fairly low, ca. 80 μV/K, compared to 110 μV/K [19] 140 μV/K [5]. This causes a low power factor, $S^2\sigma$.

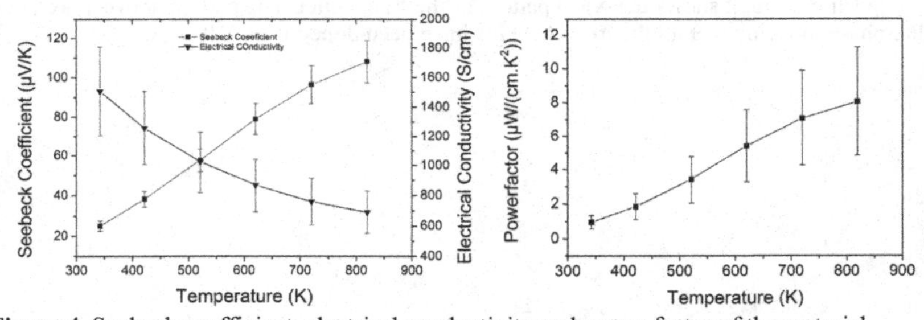


**Figure 4**. Seebeck coefficient, electrical conductivity and power factor of the material

Our aim is to integrate the fabricated material in power generation modules at intermediate temperature, to develop the module technology for skutterudites. It has been debated in the literature on the thermal stability of skutterudites and there are worries about their stability under the processing conditions for module fabrication and contacting. We therefore performed thermal stability measurements on the sample using TGA. The temperature profile is shown in Fig. 5a, where a stepwise heating and holding up to 750°C was applied to investigate the temperature stability of the skutterudite material. The thermal stability of the consolidated material was high

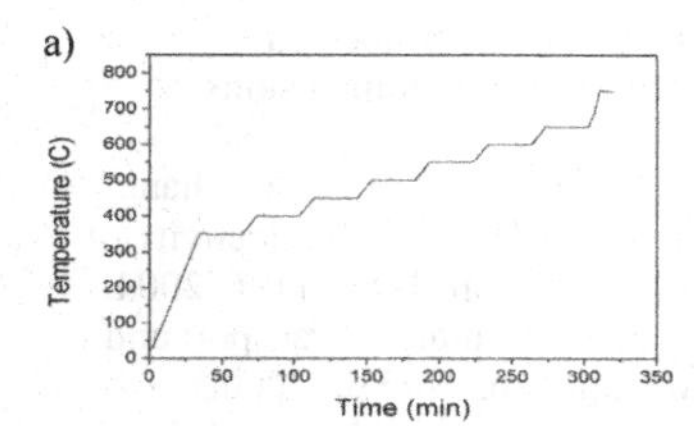


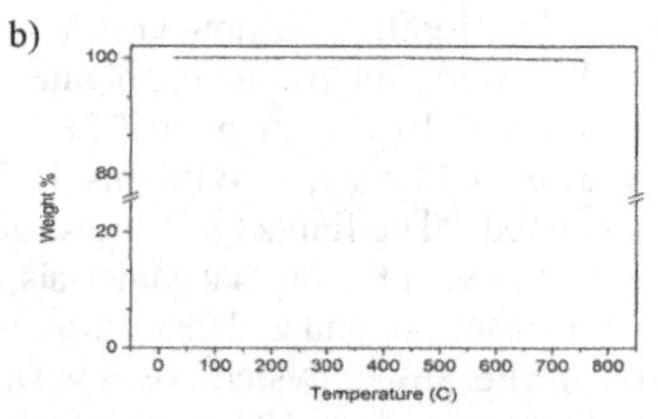


**Figure 5**. (a) Temperature profile of TGA and (b) weight profile of the skutterudite material.

enough for sudden increases for the contacting process for TE module/device fabrication. Although a very slight weight loss is visible (less than 0.05%), which correspond to oxidation and sublimation of antimony as TGA purging system is not completely sealed for leakage. Additionally, this evaporation can be passivized in the packaging processes.

## CONCLUSIONS

Large quantities of high phase purity iron doped skutterudite has been produced using chemical alloying process. The process is fast and has less impact on environment compared to other processes due to lower heat, and time consumption. In short the process involved two major steps as (i) co-precipitation of oxalate compounds in the pH region of about 2.5 to 3.5, and (ii) calcination at 350°C to have the metal oxides and reduction under hydrogen to yield the doped skutterudite material. Preliminary transport property results and of the thermal stability are promising for intermediate temperature applications.

## ACKNOWLEDGMENTS

This work has been funded by EC-FP7 program under NEXTEC project and in part by the Swedish Foundation of Strategic Research - SSF. MYT acknowledges the support from Ångpanneföreningens Forskningsstiftelse.

## REFERENCES

[1] C. Zhou, J. Sakamoto, and D. Morelli, "Low-Temperature Thermoelectric Properties of $Co_{0.9}Fe_{0.1}Sb_3$-Based Skutterudite Nanocomposites with $FeSb_2$ Nanoinclusions," Journal of Electronic Materials, vol. 40, no. 5, pp. 547–550, 2010.
[2] Y. Kawaharada, K. Kurosaki, M. Uno, and S. Yamanaka, "Thermoelectric properties of $CoSb_3$," vol. 315, pp. 193–197, 2001.
[3] Toprak, M.; Zhang, Yu; Muhammed, M.; Zakhidov, A.A.; Baughman, R.H.; Khayrullin, I., "Chemical route to nano-engineered skutterudites," Thermoelectrics, 1999. Eighteenth International Conference on , vol., no., pp.382,385, Aug. 29 1999-Sept. 2 1999
doi: 10.1109/ICT.1999.843410
[4] David. M. Rowe, "Nanostructured Skutterudite" in Thermoelectrics Handbook: Macro to Nano" 1st edition Great Britain, Taylor and Francis, 2006, chapter 41, Page 41-1, 978-0849322648

[5] C. Zhou, J. Sakamoto, D. Morelli, X. Zhou, G. Wang, and C. Uher, "Thermoelectric properties of $Co_{0.9}Fe_{0.1}Sb_3$-based skutterudite nanocomposites with $FeSb_2$ nanoinclusions," Journal of Applied Physics, vol. 109, no. 6, p. 063722, 2011.
[6] M. S. Toprak, C. Stiewe, D. Platzek, S. Williams, L. Bertini, E. Müller, C. Gatti, Y. Zhang, M. Rowe, and M. Muhammed, "The Impact of Nanostructuring on the Thermal Conductivity of Thermoelectric $CoSb_3$," Advanced Functional Materials, vol. 14, no. 12, pp. 1189–1196, 2004.
[7] J. Yang, G. Meisner, D. Morelli, and C. Uher, "Iron valence in skutterudites: Transport and magnetic properties of $Co_{1-x}Fe_xSb_3$," Physical Review B, vol. 63, no. 1, p. 014410, 2000.
[8] David. M. Rowe, "Skutterudite-based Thermoelectrics" in Thermoelectrics Handbook: Macro to Nano" 1st edition Great Britain, Taylor and Francis, 2006, chapter 34, Page 34-1, 978-0849322648
[9] K. H. Park, S. C. Ur, and I. H. Kim, "Thermoelectric Properties of Co1-xFexSb3 Prepared by Encapsulated Induction Melting," Solid State Phenomena, vol. 124–126, pp. 939–942, 2007.
[10] X. Shi, H. Kong, C.-P. Li, C. Uher, J. Yang, J. R. Salvador, H. Wang, L. Chen, and W. Zhang, "Low thermal conductivity and high thermoelectric figure of merit in n-type $Ba_xYb_yCo_4Sb_{12}$ double-filled skutterudites," Applied Physics Letters, vol. 92, no. 18, p. 182101, 2008.
[11] H. Anno, K. Matsubara, Y. Notohara, T. Sakakibara, and H. Tashiro, "Effects of doping on the transport properties of $CoSb_3$," Journal of Applied Physics, vol. 86, no. 7, p. 3780, 1999.
[12] L. Yang, H. H. Hng, D. Li, Q. Y. Yan, J. Ma, T. J. Zhu, X. B. Zhao, and H. Huang, "Thermoelectric properties of p-type $CoSb_3$ nanocomposites with dispersed $CoSb_3$ nanoparticles," Journal of Applied Physics, vol. 106, no. 1, p. 013705, 2009.
[13] X. Tang, Q. Zhang, L. Chen, T. Goto, and T. Hirai, "Synthesis and thermoelectric properties of p-type- and n-type-filled skutterudite $R_yM_xCo_{4-x}Sb_{12}$ (R:Ce,Ba,Y;M:Fe,Ni)," Journal of Applied Physics, vol. 97, no. 9, p. 093712, 2005.
[14] X. Yang, P. Zhai, L. Liu, and Q. Zhang, "Thermodynamic and mechanical properties of crystalline $CoSb_3$: A molecular dynamics simulation study," Journal of Applied Physics, vol. 109, no. 12, p. 123517, 2011.
[15] A. Zhou, L. Liu, P. Zhai, W. Zhao, and Q. Zhang, "Electronic structure and transport properties of single and double filled $CoSb_3$ with atoms Ba, Yb and In," Journal of Applied Physics, vol. 109, no. 11, p. 113723, 2011.
[16] N. Dong, X. Jia, T. C. Su, F. R. Yu, Y. J. Tian, Y. P. Jiang, L. Deng, and H. a. Ma, "HPHT synthesis and thermoelectric properties of $CoSb_3$ and $Fe_{0.6}Co_{3.4}Sb_{12}$ skutterudites," Journal of Alloys and Compounds, vol. 480, no. 2, pp. 882–884, 2009.
[17] M. D. Hornbostel, E. J. Hyer, J. Thiel, and D. C. Johnson, "Rational Synthesis of Metastable Skutterudite Compounds Using Multilayer Precursors," vol. 7863, no. 96, pp. 2665–2668, 1997.
[18] C. Zhou, D. Morelli, X. Zhou, G. Wang, and C. Uher, "Thermoelectric properties of P-type Yb-filled skutterudite $Yb_xFe_yCo_{4-y}Sb_{12}$," Intermetallics, vol. 19, no. 10, pp. 1390–1393, 2011.
[19] K. Park, S. You, S. Ur, and I. Kim, "Electronic Transport Properties of Fe-doped $CoSb_3$ Prepared by Encapsulated Induction Melting," pp. 1–4, 2006.
[20] K. H. Park, J. Il Lee, S. C. Ur, and I. H. Kim, "Thermoelectric Properties of Fe-Doped $CoSb_3$ Prepared by Encapsulated Induction Melting and Hot Pressing," Materials Science Forum, vol. 534–536, pp. 1557–1560, 2007.
[21] G. a. Slack and V. G. Tsoukala, "Some properties of semiconducting $IrSb_3$," Journal of Applied Physics, vol. 76, no. 3, p. 1665, 1994.

# Nanowires, Nanotubes, and Nanocrystals

Mater. Res. Soc. Symp. Proc. Vol. 1543 © 2013 Materials Research Society
DOI: 10.1557/opl.2013.929

# A Novel Approach to Synthesize Lanthanum Telluride Thermoelectric Thin Films in Ambient Conditions

Su (Ike) Chih Chi[1], Stephen L. Farias[1], and Robert C. Cammarata[1-2]
[1]Department of Materials Science and Engineering, Johns Hopkins University, 3400 N. Charles Street, Baltimore, Maryland 21218, USA
[2]Department of Mechanical Engineering, Johns Hopkins University, 3400 N. Charles Street, Baltimore, Maryland 21218, USA

## ABSTRACT

Rare-earth telluride compounds are characterized by their high performance thermoelectric properties that have been applied to the development of functional materials [1]. Recently, May and co-workers reported that nanostructured bulk lanthanum telluride ($La_{3-x}Te_4$, $0 \leq x \leq 1/3$) by mechanical ball-milling exceeded the figure of merit (ZT) of 1 at high temperatures near 1300K [2-3]. Since the increased thermoelectric efficiency of nanostructured materials is due to the enhancement of phonon scattering introduced by quantum confinement, thin films have also generated significant scientific and technological interest [4-6]. Here, we report on the electrodepostion of lanthanum telluride and lanthanum thin films in ionic liquids in ambient conditions. Surface morphologies varied from needle-like to granular structures and depend on deposition conditions. This novel electrochemical synthesis approach is a simple, inexpensive and laboratory-environment friendly method of synthesizing nanostructured thermoelectric materials.

## INTRODUCTION

Rare-earth elements and their telluride have attracted significant attentions because of their potential applications in thermoelectrics, electronics and optoelectronics [7]. In particular, lanthanum and lanthanum telluride alloys are of interest because lanthanum is used in large quantities in nickel metal hydride (NiMH) rechargeable batteries for hybrid automobiles. The negative electrode (cathode) in NiMH batteries is a mixture of metal hydrides, one of which is typically lanthanum hydride. The active material at the cathode is hydrogen, which is stored in the metal hydride structure [8]. A Toyota Prius battery requires 10 to 15 kg (22 to 33 lb) of lanthanum [9-10]. Lanthanum can also be used to make the infrared-absorbing glass in night vision goggles. High quality camera and telescope lenses contain lanthanum oxide ($La_2O_3$) making use of its high refractive index and low dispersion [11]. In addition, May and co-workers reported that nanostructured bulk lanthanum telluride ($La_{3-x}Te_4$, $0 \leq x \leq 1/3$) produced by mechanical ball-milling exceeded the figure of merit (ZT) of 1 at high temperatures near 1300K [2-3]. LaTe and $La_2Te_3$ are good candidates for photoconducting and photovoltaic cells applications [12]. $LaTe_2$ and $LaTe_3$ are n-type semiconductors having unique electronic properties [13].

Electrochemical deposition is a very attractive process for synthesizing thick films of compound semiconductors on metallic surfaces [14]. An advantage of this technique is that it is a simple, inexpensive, and laboratory-environment friendly synthesis methods for telluride based thermoelectric materials. Producing thick rare-earth telluride films can be very material consuming when using physical vapor deposition methods such as sputtering or evaporation, but films of several microns thickness can be produced in a few hours using electrochemical deposition [14]. In addition, variations in the deposition potential or solution concentration can provide different molar ratio of La-Te stoichiometric composition films. It is important to note that aqueous solutions are unsuitable for lanthanum alloys electrodeposition because hydrogen evolution can occur before the deposition of metals. Recently, room-temperature ionic liquids (RTILs) have attracted significant attention in the electrodeposition of rare-earth metals in ambient conditions [15]. Here we report on the electrodeposition of lanthanum and lanthanum telluride alloys in a hydrophobic ionic liquid, 1-ethyl-3-methylimidazolium bromide, which has the advantages of good cathodic stability, low hygroscopy, and low reactivity with oxygen.

## EXPERIMENTAL

Tellurium powder (Aldrich Chemistry, 99.8%) was dissolved in concentrated nitric acid to form the oxide cations, $(HTeO_2)^+$, at room temperature until no further reaction occurred. A tellurium seed layer (~ 100 nm thick) was then electrodeposited onto a silver substrate (cleaned and etched with $HNO_3$ : deionized water = 1 : 3) in a three-electrode electrochemical system designed in-house. The silver substrate also served as the working electrode in the experiment. A platinum mesh was used as the counter-electrode and an Ag/AgCl electrode was used as the reference electrode. The electrolyte contained 0.025M $(HTeO_2)^+$ in 3M $HNO_3$. Potentiastatic deposition was performed at -0.3V vs. Ag/AgCl reference electrode at room temperature.

The ionic liquid, 1-ethyl-3-methylimidazolium bromide, was prepared by firstly stirring 100 ml of 1-methylimidazole (Alfa Aesar, 99%) in a beaker at 70°C followed by slowly adding 300 ml of bromoethane (Sigma-Aldrich, reagent grade, 98%) into the beaker. The solution was then allowed to stir about 24 hours at 70°C. Ethyl ether (Fisher Scientific, anhydrous, and certified ACS) was added to wash the solution. The ionic liquid was then stir at 80°C for 5 hours to burn off any volatiles left in the solution.

Lanthanum telluride films were electrodeposited onto the tellurium seed layer in the three-electrode electrochemical system. The silver substrate was still served as the working electrode in the experiment. A platinum mesh was used as the counter electrode and an Ag/AgCl electrode was used as the reference electrode. The electrolyte contained 0.9g of lanthanum nitrate hydrate (Aldrich Chemistry, 99.9%) dissolved in 10 ml of 1-ethyl-3-methylimidazolium bromide ionic liquid and a few ml of 0.025M $(HTeO_2)^+$ in 3M $HNO_3$ depending on the La-Te molar ratio of the product as shown in table I. The potentiastatic deposition was performed at -1.2 or -1.5V vs. Ag/AgCl reference electrode at room temperature. Ar gas was bubbled into the electrolyte to remove oxygen from the solution throughout the experiment. Typical deposition duration was around 2 hours. The deposited films were rinsed in acetone and deionized water, followed by storing in vacuum. The morphology of the samples was investigated using a field emission scanning electron microscope (JEOL JSM-6700F) equipped with an energy dispersive X-ray spectrometer (EDAX) for composition analysis.

**Table I.** Electrodeposition conditions of La and La-Te alloys.

| Counter Electrode | Working Electrode | Electrolyte (in 10 ml ionic liquid) | Deposition Voltage (vs. Ag/AgCl Reference Electrode) | Product |
|---|---|---|---|---|
| Pt | Ag | 0.9g $La(NO_3)_3 \cdot 6H_2O$ | -1.5V | La |
| Pt | Ag | 0.2 ml [0.025M $(HTeO_2)^+$ in 3M $HNO_3$] + 0.9g $La(NO_3)_3 \cdot 6H_2O$ | -1.5V | LaTe |
| Pt | Ag | 0.3 ml [0.025M $(HTeO_2)^+$ in 3M $HNO_3$] + 0.9g $La(NO_3)_3 \cdot 6H_2O$ | -1.5V | $La_{3-x}Te_4$ |
| Pt | Ag | 0.2 ml [0.025M $(HTeO_2)^+$ in 3M $HNO_3$] + 0.9g $La(NO_3)_3 \cdot 6H_2O$ | -1.2V | $La_2Te_3$ |
| Pt | Ag | 1 ml [0.025M $(HTeO_2)^+$ in 3M $HNO_3$] + 0.9g $La(NO_3)_3 \cdot 6H_2O$ | -1.2V | $LaTe_2$ |
| Pt | Ag | 0.5 ml [0.025M $(HTeO_2)^+$ in 3M $HNO_3$] + 0.9g $La(NO_3)_3 \cdot 6H_2O$ | -1.5V | $LaTe_3$ |

## DISCUSSION

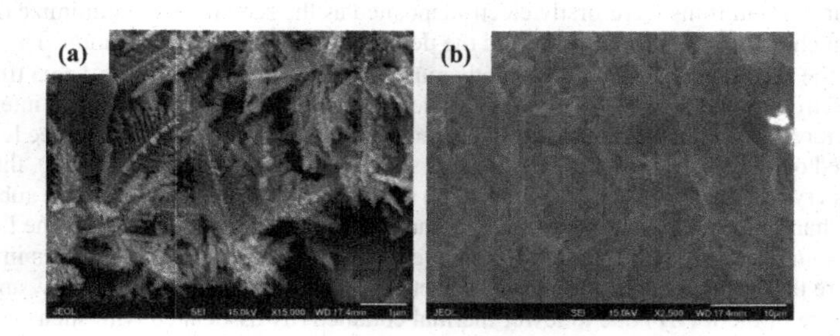

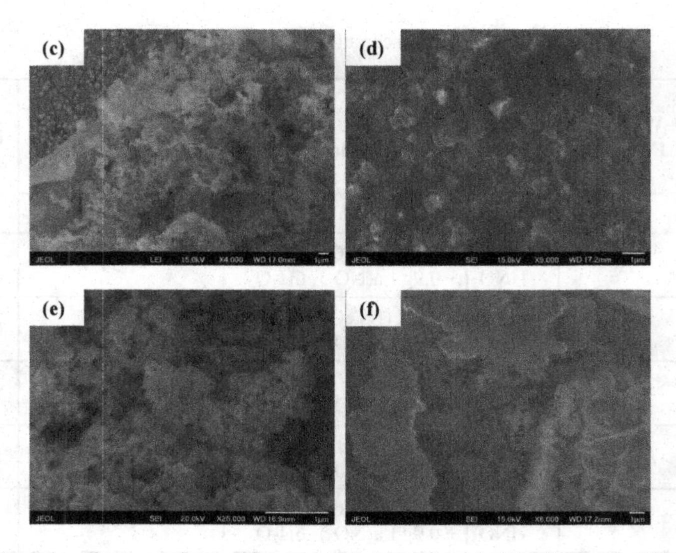


**Figure 1.** Scanning electron microscope images of electrodeposited lanthanum and lanthanum telluride films in 1-ethyl-3-methylimidazolium bromide ionic liquid at different deposition conditions (a) La; (b) LaTe; (c) $La_{3-x}Te_4$; (d) $La_2Te_3$; (e) $LaTe_2$; (f) $LaTe_3$.

Tellurium thin films were firstly electrodeposited as the seed layers to minimize the lattice mismatches between the substrate and the deposit to reduce defect formation. The thickness of the deposited lanthanum and lanthanum telluride films ranged from 1 to 5 μm. Scanning electron microscope (SEM) images of electrodeposited films display many interesting surface structures as shown in figure 1. Round needle-like structures that look like tree leaves were observed on the electrodeposited La films as shown in figure 1(a). In figure 1(b), the homogenous crystalline LaTe domains with rough surface features covered the entire substrate. On the other hand, it is noted that there are many nanostructured surface features in the $La_{3-x}Te_4$ film as shown in figure 1(c). The nanocrystalline features have many grain boundaries in the microstructure that can scatter phonons more effectively than electrons and may allow an enhanced figure of merit (ZT) due to lower thermal conductivity associated with such nanostructured materials. The chemical composition of $La_{3-x}Te_4$ films was 41-59 (La-Te) atomic % ratio, as determined from EDAX analysis as shown in figure 2. The very strong and sharp peak was identified as corresponding to the silver substrate used in this experiment. The EDAX peaks also indicate the lanthanum and lanthanum telluride samples are free of bulk oxides. The microstructures of $La_2Te_3$, $LaTe_2$, and $LaTe_3$ were composed of many nanocrystalline granular domains covered the entire substrate as respectively shown in figure 1(d)-(f).

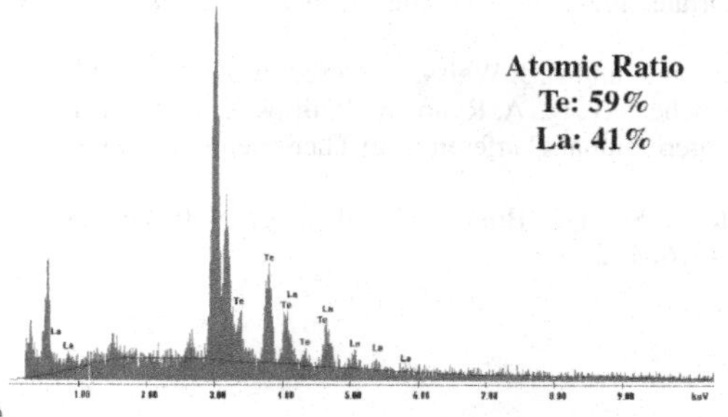


**Figure 2.** EDAX composition analysis of electrodeposited $La_{3-x}Te_4$ film.

## CONCLUSIONS

We have demonstrated the feasibility of electrodepositing pure lanthanum and lanthanum telluride compounds in ambient conditions that are free of bulk oxides. Surface morphologies varied from needle-like to granular structures depending on the deposition conditions. Many nanostructured surface features on the electrodeposited samples may enhance ZT because the appearance of a large numbers of grain boundaries can scatter phonons more effectively than electrons resulting in a lowering of the thermal conductivity.

## ACKNOWLEDGMENTS

The authors would like to thank to Ian McCue at the Johns Hopkins University for many valuable discussions on the preparation and the chemistry of ionic liquids.

## REFERENCES

1. D. M. Rowe, *CRC Handbook of Thermoelectrics*, CRC Press (1995).
2. A. May, J-P. Fleurial and G. J. Snyder, *Phys. Rev. B* **78**, 125205 (2008).
3. O. Delaire, A. F. May, M. A. McGuire, W. D. Porter, M. S. Lucas, M. B. Stone , D. L. Abernathy, V. A. Ravi, S. A. Firdosy, G. J. Snyder, *Phys. Rev. B* **80**, 184302 (2009).
4. L. D. Hicks and M. S. Dresselhaus, *Phys. Rev. B* **47**, 12727 (1993).
5. L. D. Hicks and M. S. Dresselhaus, *Phys. Rev. B* **47**, 16631 (1993).
6. M. Dresselhaus, G. Chen, M. Y. Tang, R. G. Yang, H. Lee, D. Z. Wang, Z. F. Ren, J. P. Fleurial, and P. Gogna, *Adv. Mater.* **19**, 1043-1053 (2007).
7. A. K Samal and T. Pradeep, *J. Phys. Chem. C.* **114**, 5871-5878 (2010).
8. John J. C. Kopera, "Inside the Nickel Metal Hydride Battery," Cobasys (2004).
9. S. Gorman, "As hybrid cars gobble rare metals, shortage looms," Ruters.com, Aug 31, 2009.
10. P. Bäuerlein *et al., J. of Power Sources.* **176**, 547-554 (2008).
11. Website: http://www.chemicool.com/elements/lanthanum.html.

12. G. D. Bagde, S. D. Sartale, and C. D. Lokhande, *Mater. Chem. and Phys.* **89**, 402-405 (2005).
13. T. H. Ramsey, H. Steinfink, and E. J. Weiss, *J. Appl. Phys.* **34**, 2917 (1963).
14. J.-P. Fleurial, A. Borshchevsky, M. A. Ryan, W. Phillips, E. Kolawa, T. Kacisch, R. Ewell, Proceedings of XVI International Conference on Thermoelectrics, Dresden, Germany, p. 641-645, 1997: *IEEE*
15. S. Legeai, S. Diliberto, N. Stein, C. Boulanger, J. Estager, N. Papaiconomou, and M. Draye, *Elec. Comm.* **10**, 1661-1664 (2008).

Mater. Res. Soc. Symp. Proc. Vol. 1543 © 2013 Materials Research Society
DOI: 10.1557/opl.2013.673

# Fabrication of thermally-conductive carbon nanotubes-copper oxide heterostructures

Yuan Li[1] and Nitin Chopra[1*]
[1]Metallurgical and Materials Engineering, The University of Alabama, Tuscaloosa, AL 35401 U.S.A.
*Corresponding Author E mail: nchopra@eng.ua.edu, Tel: 205-348-4153, Fax: 205-348-2164

## ABSTRACT

A complete dry processing route is developed for the fabrication of thermally-conductive carbon nanotube (CNT)-copper oxide ($CuO_x$) heterostructures. This was achieved by the deposition of copper (Cu) onto CNTs and subsequent annealing in Ar and air environment to convert the coated Cu into $CuO_x$ nanoparticles. The survivability and diameters of CNTs were studied to ensure their integrity after the multiple processing steps and annealing temperatures (400 °C). The as-produced CNTs, air/Ar-annealed CNTs, Cu-coated CNTs, and CNT-$CuO_x$ heterostructures were characterized to study their structure, phase, and morphology using microscopy, elemental analysis, X-ray diffraction, and sheet resistance. It was observed that CNTs could survive the processing conditions and became coated with $CuO_x$ nanoparticles. The sheet resistance of CNTs coated with $CuO_x$ nanoparticles was ~4 times greater than the as-produced CNTs. The Raman spectroscopy-based estimation of thermal conductivity of CNTs and CNT-$CuO_x$ heterostructures showed 2-7 times enhancement for the latter as compared to pure $CuO_x$. In conclusion, such hybrid CNT-based heterostructures are promising for applications in thermal management.

## INTRODUCTION

Carbon nanotubes (CNTs) are promising for applications in nanoelectronics, interconnects, energy, and sensors due to their remarkable properties [1-4]. In addition, CNTs exhibit superior thermal conductivities, as high as 2000 W/m-K to 6000 W/m-K [5-8], 10 times higher than copper and silver. It is of particular interest to develop novel nanocomposites or heterostructures comprised of CNTs and conventional materials such as copper (Cu) or its oxides ($CuO_x$), where the latter are known to have low thermal conductivities. These could be potentially of use as nanofillers, thermal fluid, and multifunctional interconnects [9-11]. Such nanocomposites could allow for tunability of thermal conductivity of the hybrid materials by modulating composition, structure, interface, and morphology of the heterostructured components. Here, we demonstrate a dry processing route for controlled fabrication of CNT-$CuO_x$ heterostructures. In addition, we study the thermal conductivities of these heterostructures.

## EXPERIMENT

**Materials and Methods:** Xylene and ferrocene were used as the carbon source for CNT growth and were purchased from Fisher Scientific (Pittsburgh, PA). Acetone and ethanol were purchased from VWR (Atlanta, GA). DI water (18.1 MΩ-cm) was obtained using a Barnstead International DI water system (E-pure D4641). All chemicals were used without further purification. Silicon (Si) wafers (<100>, n-type) were purchased from IWS (Colfax, CA). Dispersion of heterostructures into ethanol was carried out in a Branson 2510 Sonicator (Danbury, CT). ATC

ORION sputtering system (AJA international, Inc., North Scituate, MA) was used. Cu target (99.95%) was purchased from VWR (Atlanta, GA). Scanning Electron Microscopy (SEM) images were obtained using FE-SEM JEOL-7000 equipped with energy dispersed X-ray spectroscopy (EDX). Raman spectra were collected using Bruker Senterra system (Bruker Optics Inc. Woodlands, TX) equipped with 785 nm laser source. Numerous measurements were performed for each sample.

**Preparation of CNT-$CuO_x$ heterostructures:** CNTs were synthesized in a floating-catalyst chemical vapor deposition (CVD) method using xylene as the carbon source and ferrocene as the catalyst [12]. The as-produced CNTs were dispersed in ethanol to form a suspension (~1 g/L). Approximately, 1 mL of this suspension was drop-casted on the oxidized silicon wafer (2 cm×2 cm) and dried in a vacuum oven at ~60 °C. The resulting substrate had a ~1 μm-thick film of CNTs and was inserted in a sputtering chamber, where a ~2.5 nm thick Cu film was sputtered on the CNTs at 100 W and $2.5 \times 10^{-3}$ Torr for 15 s. These Cu-coated CNT films were annealed (dewetted) at 400 °C for 1 h in air or Ar to result in CNT-$CuO_x$ heterostructures.

**Sheet resistivity:** Sheet resistance of the CNT-based nano-composites was measured using JANDEL RM3 four-point probe system. Forward and reverse resistance was measured in a current range of 10 μA to 10 mA. The sheet resistance values were recorded at the current (or voltage) and the measurements were repeated for 5 times.

**Thermal conductivity:** Thermal conductivity of the as-produced CNTs and CNT-$CuO_x$ heterostructures was measured using Raman spectroscopy method [13-15]. Specifically, G-band peak locations in the Raman spectra were studied as a function of laser powers (Q, 1 mW, 10 mW, 25 mW, and 50 mW) and substrate temperatures (from 25 °C to 160 °C). The measurements were repeated several times. Experimentally evaluating G-band shift vs. Q and vs. T resulted in a linear relationship between temperature and laser power as follows:

$$T = mQ + n \qquad (1)$$

where m and n are constants. The thermal conductivities were estimated using [14,15]:

$$K = \frac{S}{2\pi tm} \qquad (2)$$

where $S$ is a constant depending on $r_0$ and $R$. $R$ is the distance from the center of the laser spot uptil the room temperature location on the substrate while $r_0$ is the laser diameter. $S$ in our study was obtained to be ~3.425.

## RESULTS AND DISCUSSION

Figure 1 illustrates the approach for the preparation of CNT-$CuO_x$ heterostructures. The as-produced CNTs were annealed in air at 400 °C for 1 h to ensure their survivability under the conditions for heterostructure fabrication. Figure 2 shows the SEM images of as-produced CNTs (sample #1), air-annealed CNTs (sample #2), Cu-coated CNTs (sample #3), and CNT-$CuO_x$ heterostructures (sample #4, 5, and 6). The sample numbers (sample #1-6) are specified in figure 1. CNTs successfully survived the air-annealing process at 400 °C (Figure 2B) [12]. Cu sputter coating on CNTs resulted in uniform and polycrystalline film but due to the line-of-sight sputter deposition process, only the regions of CNTs facing the incident depositing material were coated (shown by arrows in figure 2C). These samples were annealed in three different ways: a) Annealing in Ar (sample #5), b) annealing in Ar followed by complete oxidation in air-annealing (sample #6), and c) direct air-annealing without the Ar annealing step (sample #4). It was observed (Figure 2D) that direct air-annealing of Cu-coated CNTs resulted in the oxide film rather than nanoparticle morphology of $CuO_x$. This oxide film was observed to be uniformly coating CNTs and could be attributed to surface migration of oxidized Cu during the annealing process. To prevent oxide film formation, dewetting of Cu coating in Ar was performed and followed by air-annealing to result in CNT-$CuO_x$ nanoparticles heterostructures (Figure 2E and F). It was observed that Ar annealing resulted in nanoparticles but due to air exposure, it led to the oxidation of nanoparticles in these samples (Figure 2I). The formation of nanoparticles for sample #5 could be attributed to greater self-diffusivity of Cu as compared to diffusivity of Cu in $CuO_x$ [16, 17], where the former facilitated dewetting of Cu films. The composition of the sample #3-6 was further confirmed using EDS and XRD spectra as shown in Figure 2H and I. XRD for sample #3 showed Cu and $CuO_x$ peaks, where the latter could be attributed to air oxidation. After air-annealing (sample #4) of Cu-coated CNTs, no pure Cu peaks were observed. Annealing Cu-coated samples in Ar environment (sample #5) resulted $Cu_2O$ peaks due to air exposure and for sample #6 (after air annealing of sample #5), $Cu_2O$ oxidized to CuO.

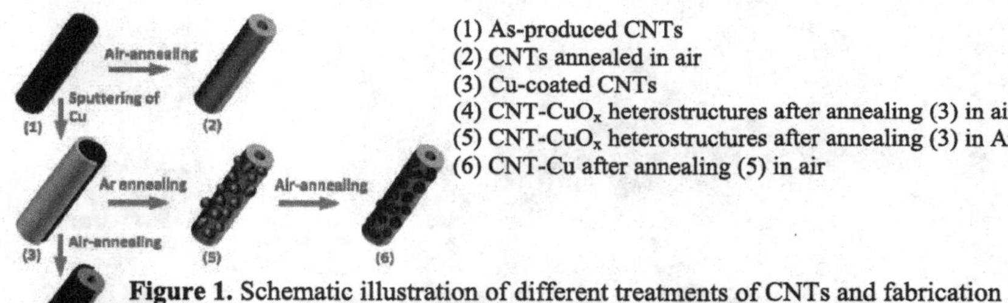


**Figure 1.** Schematic illustration of different treatments of CNTs and fabrication of CNT-$CuO_x$ heterostructures by combining Cu sputter coating with annealing process. The numbers indicate the sample numbers at each stage. Note: For any annealing step, annealing temperature selected was ~400 °C and for 1 h duration.

Further analysis of the sizes indicated the following diameter trends for CNTs and CNT-$CuO_x$ heterostructures: Cu-coated CNTs (sample #3) > as-produced CNTs (sample #1) > air-annealed CNTs (sample #2) > Ar annealed Cu-coated CNTs (sample #5) > air-annealed Cu-coated CNTs (sample #4) ~ air-annealed sample #5 (sample #6). The reduction in diameter of sample #2 is because of air oxidation, decomposition, and etching of CNTs [12]. Annealing in Ar environment prevented damage to CNTs but it has been observed that metal nanoparticles can result in CNT etching in such annealing conditions [12], which is the reason for the lower CNT diameters for sample #5 as compared to samples #1 and 3. Knowing that the density of $CuO_x$ (~6 $g/cm^3$) is lower than pure Cu (8.93 $g/cm^3$), it is expected that Cu-coated CNTs should have lower diameters than air-annealed sample #4. However, opposite to this is observed, which could be attributed to the uniform re-distribution of $CuO_x$ around the CNTs during air-annealing. In addition, etching of exposed CNTs during air-annealing could not be ruled out. This can also explain the observed diameters of sample #6, which was similar to sample #4. It was observed that $CuO_x$ nanoparticles in sample #5 were lower in diameters as compared to $CuO_x$ nanoparticles in sample #6. This could be attributed to complete oxidation of nanoparticles in sample #6 and at the same time aggregation, surface migration, and Ostwald's ripening effects dominated [18].

Since the CNTs were treated in different environment or coated with Cu or $CuO_x$, the sheet resistances of the samples were measured. It was observed that Cu-coated CNTs (sample #3) had the lowest resistance, due to the presence of Cu. This was comparable with the sheet resistance of the as-produced CNTs (sample #1). However, no significant change in sheet resistance was observed for Cu-coated CNTs after Ar annealing. The as-produced CNTs after air annealing (sample #2) resulted in 2 times increase in sheet resistance due to oxidation and damage to the CNT structures. Similar increase in air-annealed Cu-coated CNTs (sample #4) was observed and apart from CNT structural damage, this could be due to the formation of $CuO_x$, which is electrically inferior to pure Cu. The maximum increase (4 times as compared to as-produced CNTs) in sheet resistance was observed for sample #6, where the Ar-annealed Cu-coated CNTs were completely oxidized in air. This increase could be attributed to the greater extent of structural damage and $CuO_x$ formation after the annealing process.

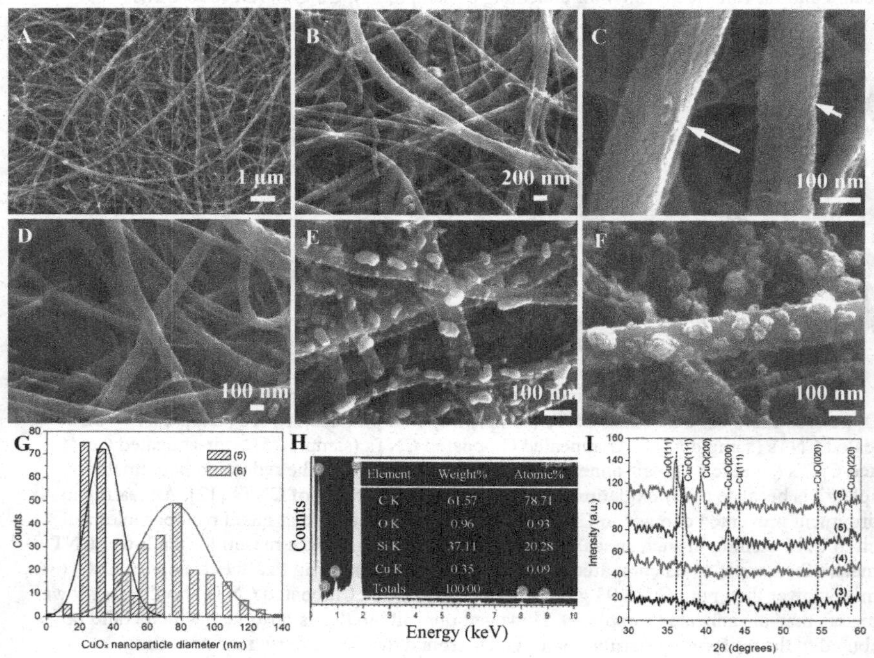


**Figure 2**. SEM images of (A) as-produced CNTs (sample #1), (B) air-annealed CNTs (sample #2), (C) Cu-coated CNTs (sample #3), (D) air-annealed CNT-$CuO_x$ heterostructures (sample #4), (E) Ar annealed CNT-$CuO_x$ heterostructures (sample #5), and (F) air-annealed sample #5 (sample #6). (G) Diameter distribution of $CuO_x$ nanoparticles on CNTs (sample #5 and 6). (H) EDS spectra (sample #6). (I) XRD spectra of CNT-based heterostructures (JCPDS for pure Cu, $Cu_2O$, and CuO are 04-0836, 35-1091, and 45-0937, respectively.)

Finally, Raman spectroscopy was utilized to study the thermal conductivity of the samples. Raman spectra for all the samples are shown in figure 3A. It can be observed that all samples exhibited characteristic peaks of CNTs: D band (around 1310 $cm^{-1}$), G band (around 1590 $cm^{-1}$), and 2D band (around 2600 $cm^{-1}$). The presence of 2D band suggests second order phonon boundary scattering and presence of defects within the CNTs. The linear trends were observed for G band shifts (Figure 3B and C) vs. Q and T. Thermal conductivities of the samples (#1, 2, 5, and 6) were estimated using equations (1) and (2). The obtained thermal conductivities are shown in Table 1. As-produced CNTs showed the highest thermal conductivity of ~ 91.56 W/m-K and this value decreased to 35.51 W/m-K (39% decrement) for air-annealed CNTs due to creation of defects and oxidation of CNTs. The defect or structural damage in CNTs could have led to significant phonon scattering resulting in lowering of thermal conductivity. It was observed that Cu-coated CNTs after Ar annealing and air annealing resulted in nearly the same thermal conductivities (~ 41 W/m-K) but these values were much greater than (2-7 times) pure $CuO_x$ [19,20]. This observation indicates significant enhancement of thermal conductivity of $CuO_x$ due to the loading onto CNTs. However, the thermal conductivities of CNT-$CuO_x$ heterostructures were lower than as-produced CNTs due to processing-induced defects in CNTs.

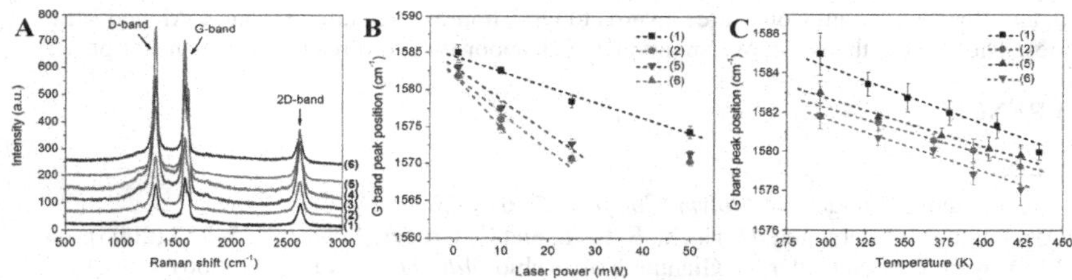


**Figure 3.** (A) Raman spectra of sample #1-6, G band peak position vs. (B) Q, (C) T.

**Table 1.** Diameters of CNTs, CNT-$CuO_x$ heterostructures, and $CuO_x$ nanoparticles as well as sheet resistivity and thermal conductivity of various samples.

| Sample # | 1 | 2 | 3 | 4 | 5 | 6 |
|---|---|---|---|---|---|---|
| **$CuO_x$ nanoparticle particle size (nm)** | --- | --- | --- | --- | **34.1±10.3** | **75.0±20.2** |
| **Nanotube diameters (nm)** | **82.2±33.1** | **75.5±26.2** | **87.2+31.7** | **65.7±27.9** | **73.9±36.7** | **66.4±24.0** |
| **Sheet resistance (Ω/□)** | **42.4±3.9** | **84.1±12.6** | **39.7±6.2** | **84.3±9.9** | **44.1±3.1** | **171.8±11.0** |
| **Thermal conductivity (W/m-K)** | **91.6** | **35.5** | --- | --- | **41.1** | **41.2** |

## CONCLUSIONS

The fabrication of CNT-$CuO_x$ heterostructures is demonstrated by combining sputter deposition and annealing processes. CVD growth CNTs were studied for their integrity and change in diameters during the fabrication, especially, during the annealing process at various stages of fabrication. It was observed that CNTs could successfully survive the multiple processing steps with some extent of structural damage/diameter reduction during sputter deposition of Cu, Ar annealing, and air annealing steps. Different morphologies of $CuO_x$ was observed under different conditions, for example, air annealing of Cu-coated CNTs resulted in a $CuO_x$ film formation and Ar annealing followed by air-annealing resulted in $CuO_x$ nanoparticle coating on CNTs. SEM, EDS, and XRD were utilized to study the morphology, phases, and structure of the hybrid heterostructures. Finally, thermal conductivity of CNTs, air-annealed CNTs, and CNT-$CuO_x$ nanoparticle heterostructures was evaluated using Raman spectroscopy method, where the shift in the G band position as a function of laser power and substrate temperature was studied. It was further observed that heterostructuring of CNTs with $CuO_x$ nanoparticles resulted in significant enhancement (2-7 times) of the thermal conductivity of the heterostructures as compared to pure $CuO_x$. This study provides promising results in the development of novel thermally conductive heterostructures useful for applications in thermal management. In addition, it also reveals that CNTs can survive harsh conditions (high temperatures and oxidizing environments) as well as multiple processing steps necessary to fabricate thermally-conductive heterostructures.

## ACKNOWLEDGMENTS

This work was funded by National Science Foundation (No. 0925445), 2012 NSF-EPSCoR RII award, and Research Grant Committee awards to Dr. Chopra. The authors thank CAF and MFF facilities. The authors thank Dr. W. Shi and Dr. S. Kapoor for proof reading the manuscript.

## REFERENCES

---

**1.** O. Breuer, and U. Sundararaj, *Polym. Compos.* **25**, 630 (2004).
**2.** E. Artukovic, M. Kaempgen, D. Hecht, S. Roth, and G. Gruner, *Nano Lett.* **5**, 757 (2005).
**3.** P. M. Ajayan, L. S. Schadler, C. Giannaris, A. Rubio, *Adv. Mater.* **12**, 750 (2000).
**4.** A. K. Geim, and K. S. Novoselov, *Nature Mater.* **6**, 183 (2007).
**5.** M. Fujii, X. Zhang, H. Xie, H. Ago, K. Takahashi, T. Ikuta, H. Abe, and T. Shimizu, *Phys. Rev. Lett.* **95**, 065502 (2005).
**6.** E. Pop, D. Mann, Q. Wang, K. Goodson, and H. Dai, *Nano Lett.* **6**, 96 (2006).
**7.** C. H. Liu and S. S. Fan, *Appl. Phys. Lett.* **86**, 123106 (2005).
**8.** S. Shenogin, A. Bodapati, L. Xue, and R. Ozisik, *Appl. Phys. Lett.* **85**, 2229 (2004).
**9.** P. K. Namburu, P. K. Praveen, and M. Debasmita, *Exp. Therm. Fluid Sci.* **32**, 397 (2007).
**10.** N. Chopra, B. Hu, and B.J. Hinds, J. Mater. Res. **22**, 2691 (2007).
**11.** S. P. Jang, and U. C. Stephen, *Appl. Therm. Eng.* **26**, 2457 (2006).
**12.** N. Chopra, W. Shi, and A. Bansal, *Carbon*, **49**, 3645 (2011).
**13.** A. A. Balandin, S. Ghosh, W. Bao, I. Calizo, D. Teweldebrhan, F. Miao, and C. N. Lau, *Nano Lett.* **8**, 902 (**2008**).
**14.** W. Cai, A. L. Moore, Y. Zhu, X. Li, S. Chen, and L. Shi, *Nano Lett.* **10**, 1645 (2010).
**15.** D. Kim, L. Zhu, C. Han, J. Kim, and S. Baik, *Langmuir* **27**, 14532 (2011).
**16.** D. B. Butrymowicz, J. R. Manning, and M.l E. Read, *J. Phys. Chem. Ref. Data* **2**, 643 (1973).
**17.** G. Yakunin. *J. Mater. Chem.* **7** 2085 (1997).
**18.** W. Shi, and N. Chopra, *J. Nanopart. Res.* **13**, 851 (2011).
**19.** K. Kwak, and C. Kim, *Korea-Aust. Rheol. J.* **17**, 35 (2005).
**20.** S. Schreck and M. Rohde. In *SPIE LASE: Lasers and Applications in Science and Engineering*, 72020A (2009).

**Mater. Res. Soc. Symp. Proc. Vol. 1543 © 2013 Materials Research Society**
**DOI: 10.1557/opl.2013.988**

# Enhanced thermopower of GaN nanowires with transitional metal impurities

G. A. Nemnes[1], Camelia Visan[2], T. L. Mitran[1], Adela Nicolaev[1], L. Ion[1] and S. Antohe[1]
[1]University of Bucharest, Faculty of Physics, ''Materials and Devices for Electronics and Optoelectronics'' Research Center, P.O. Box MG-11, 077125 Magurele-Ilfov, Romania
[2]''Horia Hulubei'' National Institute for Physics and Nuclear Engineering (IFIN-HH), 077126 Magurele-Ilfov, Romania

## ABSTRACT

The thermopower properties of GaN nanowires with transitional metal impurities are investigated in the framework of constrained spin density functional theory (DFT) calculations. The nanowires are connected to nanoscopic Al[111] electrodes, which ensure a natural coupling to the wurtzite structure of the nanowires. We investigate the thermoelectric properties comparatively for the pristine GaN nanowire and the system with one Mn adatom. Our study points out the predicted qualitative behavior for systems with a peak in the total transmission, as well as the sign change in the thermopower. For the system with the magnetic impurity we find an enhanced conductance, thermopower and figure of merit. The detectable spin current polarization suggests the device structure may be also used in low temperature sensing applications.

## INTRODUCTION

The ongoing efforts for producing efficient thermoelectric devices currently include the investigation of nanowires [1] and nanotubes [2] or even atomic chains [3] and single molecules [4] contacted to nano- or bulk electrodes. Besides generating thermoelectricity, nanostructures of this type constitute the main building blocks of high performance temperature sensors and cooling devices.

The focus on atomic sized thermoelectric devices is not only supported by the benefits of scaling, but also by the enhanced thermopower (Seebeck coefficient), which arises from typically sharp variations of the device conductance. According to the Cuttler-Mott formula [5-7],

$$S^{CM}(\mu, T) = -\frac{\pi^2 k_B^2 T}{3e}\frac{\partial \ln G}{\partial \mu}, \quad (1)$$

the Seebeck coefficient, $S^{CM}$, depends on the derivative of the logarithmic conductance, $\ln G$, with respect to the chemical potential, $\mu$. Rapid variations of the conductance were found in nanowire systems with attractive potential impurities [8]. The transmission function presents a series of sharp dips in front of each plateau, which gives a relatively high thermopower. Depending on the position of the chemical potential, by raising the temperature it is also possible to obtain a change in sign for the Seebeck coefficient.

The thermopower of atomic sized wurtzite AlN nanowires, with Al[111] bulk contacts, has been investigated recently [9]. The group-III nitrides are large bandgap semiconductors.

However, in the case of atomic sized nanowires it was established that reasonable conduction can be achieved due to the appearance of surface states. Tunning and controlling the position of the surface states is one essential factor in establishing the condition for a high performance thermoelectric device.

In the present study we analyze thermoelectric effects in GaN nanowires with nanoscopic Al contacts, taking into account the influence of transitional metal impurities. The GaN has a smaller bandgap than AlN, which renders better conduction at low temperatures. The Mn adatoms placed on the nanowire surface enhance both the conduction and the thermopower of the considered device. A brief discussion about the polarization of the spin currents is also presented.

## THEORY

The analyzed structures consist of GaN nanowires contacted to Al[111] nanocontacts, as indicated in Fig. 1 (a) and (b). The wire segments consists of 2 unit cells, each containing 12 atoms (6 Ga and 6 N atoms), and one extra layer of Ga, which connects to the Al electrode.The A-B-C type stacking of fcc Al[111] ensures a natural coupling to the wurtzite structure of the wire, since the adjacent layers at the interface are both hexagonal, with a minimal lattice mismatch [10].

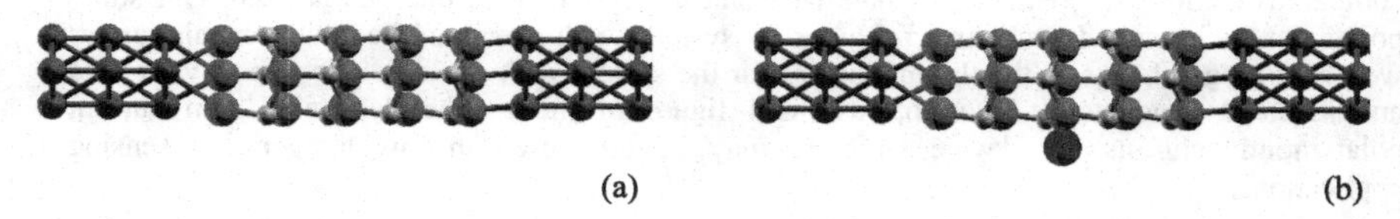

(a) (b)

**Figure 1.** (a) Pristine wurtzite GaN nanowire connected to Al[111] nanocontacts. (b) GaN nanowire with one Mn adatom.

Structural relaxations were performed in the framework of spin constrained DFT calculations, implemented by SIESTA [11]. For the exchange-correlation potential, we used the local spin density approximation (LSDA), in the parametrization proposed by Ceperley and Alder. The initial configuration for the GaN is set using the experimental values of the lattice parameters, $a_0^{GaN} = 3.189A^{\circ}$, $c_0^{GaN} = 5.185A^{\circ}$. Following relaxations, the GaN nanowires retain a slightly contracted wurtzite configuration, with $a^{GaN} = 2.89A^{\circ}$, $c^{GaN} = 5.1A^{\circ}$. The relaxed parameters are obtained using a Monkhorst-Pack sampling scheme of $1 \times 1 \times 5$ points and maximum allowed forces of $0.04eV/A^{\circ}$. The lattice constant of the fcc Al[111] nanoscopic contacts is $a^{Al} = 4.08A^{\circ}$, which implies a hexagonal lattice parameter in the A-B-C layers of $a_h^{Al} = \frac{a^{Al}}{\sqrt{2}} = 2.86A^{\circ}$, which is very close to $a^{GaN}$.

The transport is evaluated within the non-equilibrium Green's functions formalism (NEGF), implemented by the additional package TRANSIESTA [12]. The spin-dependent transmission functions, $T_{\uparrow}$ and $T_{\downarrow}$, are obtained in the small (linear) bias regime. Next, the linear response functions can be extracted,

$$L_m = e^{-m} \int_{-\infty}^{\infty} dE \, (T_{\uparrow} + T_{\downarrow})(E) \times (E - \mu)^m \left( -\frac{\partial f_{FD}(E, \mu, T)}{\partial E} \right), \quad (2)$$

where $(T_\uparrow + T_\downarrow)(E)$ is the total transmission function and $f_{FD}(E, \mu; T)$ is the Fermi-Dirac distribution. The linear regime conductance $G(\mu; T)$, the Seebeck coefficient $S(\mu; T)$ and the thermoelectric figure of merit $ZT$ may be written as:

$$G(\mu; T) = \frac{2e^2}{h} L_0(\mu; T) \quad (3)$$

$$S(\mu; T) = \frac{1}{T} \frac{L_1(\mu; T)}{L_0(\mu; T)} \quad (4)$$

$$ZT = \frac{1}{\frac{L_0 L_2}{L_1^2} - 1}. \quad (5)$$

## DISCUSSION

We start our analysis by looking at the transmission functions. In Fig. 2 (a) and (b) are plotted the spin dependent transmission functions for the pristine GaN nanowire and the system with one additional Mn impurity, respectively. In the case of the non-magnetic system indicated in Fig. 2(a) one obtains $T_\uparrow = T_\downarrow$. Upon inserting the Mn adatom, which has an incomplete d-shell and therefore a net magnetic moment, the transmissions for the two spin components become different. In our simulation, the magnetic moment of the magnetic impurity was set to *up*-spin.

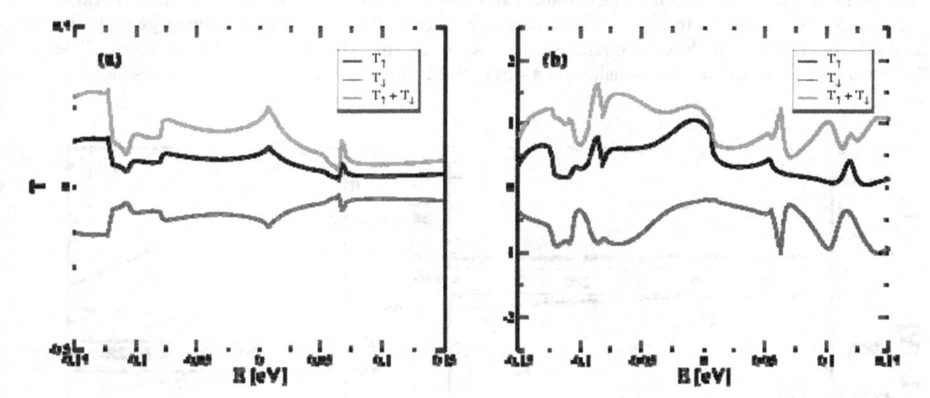


**Figure 2.** Spin dependent transmission functions for the pristine GaN nanowire with Al nanocontacts (a) and for the system with one Mn adatom. The total spin transmission is also indicated. The chemical potential is marked by vertical dotted lines ($\mu = 0$).

The Al[111] nanocontacts are non-magnetic and they inject a non-polarized spin current. As indicated in other recent studies, the presence of transitional metal impurities introduces a significant spin current polarization [10,13]. However, more importantly in the present study is the enhancement of the total transmission in the system with the magnetic impurity. One can also note that due to the relatively short GaN nanowire segment, there is a strong coupling between the two contacts mediated by the surface states. The consequence is that the two analyzed systems have a considerable conduction around the Fermi energy.

Once the transmissions are determined, one can evaluate the linear response functions and subsequently one can obtain the thermoelectric quantities of interest. The functions $L_{m}$, as well as the thermopower and the figure of merit indicated in Fig. 3 are evaluated up to room temperature. The first linear response function, $L_0$, evaluated at temperature T=0K represents the total transmission at the Fermi energy and one can see from Fig. 3 (a) that the conductance is much larger for the system with one Mn impurity on the entire temperature range considered. The functions $L_1$ become negative due to the overall decreasing transmission with energy for both systems, and the transmission in the region with $(E-\mu) < 0$ has a larger influence. In the case of $L_2$ functions, the factor $(E-\mu)^2$ is always positive. The Seebeck coefficient has a qualitatively different behavior for the two systems, which is consistent with the data regarding the total transmission function presented in Fig. 2. One one hand, for the pristine GaN nanowire, the transmission function is increasing with energy around the Fermi level up to an energy $E \cong 6meV$. This maximum is visible in the thermopower, in Fig. 3(d), as well as in the subplot (e), where the figure of merit is indicated. A similar behavior was indicated in Ref [8], where the case of a *dip* in the transmission function was analyzed. However it was predicted that in case of a *peak* in the total transmission – which is the case in the present study, a mirrored symmetry is found: a maximum in the Seebeck coefficient is found instead of a minimum and in both cases there is a change in sign for *S*. On the other hand, for the system with a magnetic impurity there is a pronounced decrease in total transmission in Fig. 2(b). From this follows the large negative Seebeck coefficient at low temperatures as well as the enhanced figure of merit. The thermopower obtained for the considered atomic sized devices is in the range of a few tens of $\mu V/K$.

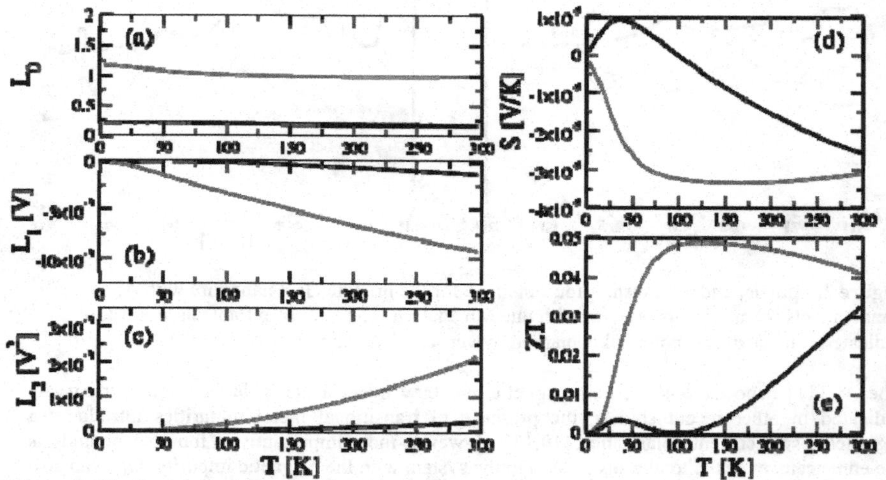


**Figure 3.** Linear response functions $L_{m}$ (a-c), thermopower (d) and figure of merit (e), for pristine GaN nanowire (black) and for the system with one Mn impurity (red).

It is worth mentioning that the atomic sized device with a transitional metal impurity produces a spin polarized current. It follows that the applied temperature difference can be measured from the polarization of the spin current, which suggests the possibility to consider applications of spintronic devices for high sensitivity temperature sensors.

## CONCLUSIONS

Thermoelectric properties of GaN nanowires with transitional metal impurities, connected to Al[111] nanocontacts were investigated using spin constrained DFT calculations. It was established that the pristine GaN nanowire exhibits a peak-like feature in the total transmission, which leads to a maximum in the Seebeck coefficient as a function of temperature, followed by a sign change. This matches the behavior predicted in the literature. Furthermore, the system with one Mn adatom indicates a significantly larger transmission, which follows from adjusting the surface states. This also leads to a larger thermopower and figure of merit. In the end it is mentioned the possibility of using the device structure in applications for highly sensitive temperature sensors.

## ACKNOWLEDGMENTS

This work was supported by the Romanian National Authority for Scientific Research, CNCS-UEFISCDI, Projects nos. PN-II-RU-PD-2011-3-0044, PN-II-ID-PCE-2011-3-0960, PN 09370104/2013 and by the European Commission under EU FP7 Project HP-SEE (under Contract no. 261499).

## REFERENCES

1. T. Markussen, A.-P. Jauho, M. Brandbyge, Phys. Rev. B **79**, 035415 (2009)
2. Y.-F. Li, B.-R. Li, H.-L. Zhang, J. Phys.: Condens. Matter **20**, 415207 (2008)
3. X.Q. Shi, Z.X. Dai, X.H. Zheng, Z. Zeng, J. Chem. Phys. B **110**, 16902 (2006)
4. Z.X. Dai, X.H. Zheng, X.Q. Shi, Z. Zeng, Phys. Rev. B **72**, 205408 (2005)
5. M. Cutler, N.F. Mott, Phys. Rev. **181** (3), 1336 (1969)
6. U. Sivan, Y. Imry, Phys. Rev. B **33** (1), 551 (1986)
7. P. Streda, J. Phys.: Condens. Matter **1**, 1025 (1989)
8. G.A. Nemnes, L. Ion, S. Antohe, Physica E **42**, 1613 (2010)
9. G.A. Nemnes, C. Visan, S. Antohe, Physica E **44**, 1092 (2012)
10. G.A. Nemnes, J. Nanomater. 408475 (2013)
11. J. M. Soler, E. Artacho, J. D. Gale et al., J. Phys.: Condens. Matt. **14**, 2745 (2002)
12. M. Brandbyge, J. L. Mozos et al., Phys. Rev. B **65**, 165401 (2002)
13. G.A. Nemnes, J. Nanomater. 748639 (2012)

**Mater. Res. Soc. Symp. Proc. Vol. 1543 © 2013 Materials Research Society**
**DOI: 10.1557/opl.2013.933**

# Growth of Polycrystalline Indium Phosphide Nanowires on Copper

Kate J. Norris[1,2], Junce Zhang[1,2], David M. Fryauf[1,2], Elane Coleman[3], Gary S. Tompa[3], and Nobuhiko P. Kobayashi[1,2]
[1]Baskin School Of Engineering, University of California, Santa Cruz, Santa Cruz, CA, United States.
[2]Nanostructured Energy Conversion Technology and Research (NECTAR), Advanced Studies Laboratories, Univ. of California Santa Cruz – NASA Ames Research Center, Moffett Field, CA, United States.
[3]Structured Materials Industries, Inc., Piscataway, NJ, United States.

## ABSTRACT

Our nation discards more than 50% of the total input energy as waste heat in various industrial processes such as metal refining, heat engines, and cooling. If we could harness a small fraction of the waste heat through the use of thermoelectric (TE) devices while satisfying the economic demands of cost versus performance, then TE power generation could bring substantial positive impacts to our society in the forms of reduced carbon emissions and additional energy. To increase the unit-less figure of merit, ZT, single-crystal semiconductor nanowires have been extensively studied as a building block for advanced TE devices because of their predicted large reduction in thermal conductivity and large increase in power factor. In contrast, polycrystalline bulk semiconductors also indicate their potential in improving overall efficiency of thermal-to-electric conversion despite their large number of grain boundaries. To further our goal of developing practical and economical TE devices, we designed a material platform that combines nanowires and polycrystalline semiconductors which are integrated on a metallic surface. We will assess the potential of polycrystalline group III-V compound semiconductor nanowires grown on low-cost copper sheets that have ideal electrical/thermal properties for TE devices. We chose indium phosphide (InP) from group III-V compound semiconductors because of its inherent characteristics of having low surface states density in comparison to others, which is expected to be important for polycrystalline nanowires that contain numerous grain boundaries. Using metal organic chemical vapor deposition (MOCVD) polycrystalline InP nanowires were grown in three-dimensional networks in which electrical charges and heat travel under the influence of their characteristic scattering mechanisms over a distance much longer than the mean length of the constituent nanowires. We studied the growth mechanisms of polycrystalline InP nanowires on copper surfaces by analyzing their chemical, optical, and structural properties in comparison to those of single-crystal InP nanowires formed on single-crystal surfaces. We also assessed the potential of polycrystalline InP nanowires on copper surfaces as a TE material by modeling based on finite-element analysis to obtain physical insights of three-dimensional networks made of polycrystalline InP nanowires. Our discussion will focus on the synthesis of polycrystalline InP nanowires on copper surfaces and structural properties of the nanowires analyzed by transmission electron microscopy that provides insight into possible nucleation mechanisms, growth mechanisms, and the nature of grain boundaries of the nanowires.

## INTRODUCTION

Study of group III-V compound semiconductor nanowires have been almost exclusively focused on single-crystal nanowires grown on single-crystal substrates with few exceptions.

Polycrystalline nanowires have previously been reported [1-3]. The growth of III-V nanowires with copper has only recently been possible, but copper is the seed and not the substrate [4]. There are many possible applications for which the growth of semiconductor nanowires on metal would be advantageous. Metal is conductive electrically and thermally as well as low cost [5]. Therefore polycrystalline nanowires could be advantageous for thermoelectric devices by combining the benefits from polycrystalline films and nanowires to achieve maximum phonon scattering through structuring.

We are pleased to report the growth of polycrystalline indium phosphide nanowires on flexible copper foil. Thermoelectric devices could benefit from both unusual aspects of this research. The ability to grow semiconductor nanowires directly on metallic substrates is possible but challenging [6]. While many have been looking toward polycrystalline materials for thermoelectric applications, this allows for both the advantages of nanowires and polycrystallinity while eliminating the problem of developing thermoelectric structures on poor electrical contact materials. The possible use of this research for a new base of a thermoelectric platform could allow for this technology to meet the cost performance demands of the market.

## EXPERIMENT

In this study, we used polycrystalline copper (Cu) foil for the growth of indium phosphide. Two types of copper substrates were prepared. The Cu foil not treated with carbon was cleaned with acetone, isopropanol, methanol, and rinsed in DI water. This was followed by the application of catalysts; 10 nm colloidal gold particles were drop casted 3 times to increase density. The second Cu foil was cleaned with acetic acid then rinsed in DI water, and dried in air prior to carbon deposition. The carbon deposition was conducted by annealing the Cu foil in hydrogen at 990□C and 270 mTorr for 20 minutes followed by flow of methane/hydrogen mixture for 10 minutes, also at 990□C and 270 mTorr. Following deposition the samples were allowed to cool in hydrogen then purged with argon before removal from the reactor. Subsequently, 5 nm, 10 nm, and 50 nm gold colloids were drop casted 3 times for thorough coverage. Metal organic chemical vapor deposition (MOCVD) was used to grow indium phosphide (InP) on the prepared copper substrates. Both substrates had similar growth parameters; 20 minutes at 300 Torr, 550°C, and a 4.3 V/III molar ratio. The precursors for the growths of these two InP samples were ditertiary butyl phosphine (DTBP) and trimethylindium (TMIn). The grown InP nanowires were then characterized by scanning electron microscope (SEM), energy dispersive x-ray spectroscope (EDS), and transmission electron microscope (TEM) to assess differences in their growth mechanisms that depend on the presence of the carbon layer.

## DISCUSSION

The two InP samples grown on the copper substrates prepared with and without carbon treatment showed substantial difference in their growth morphology. InP grew as a polycrystalline thin film on the copper substrate prepared without a carbon layer while InP grew as polycrystalline nanowires on those prepared with a carbon layer. Figure 1a shows the SEM images taken on the copper substrate without carbon post growth, indicating that a film, poly-crystalline in nature, developed. This conclusion was made due to the crystal faces we see emerging on the surface of the film. However, this was not confirmed through X-ray diffraction, and therefore, we cannot show this quantitatively. The EDS confirms that the grown thin film is composed of indium and phosphorus with an atomic percentage approximation of ~1:1 In:P, which suggests that the film is composed of indium phosphide.

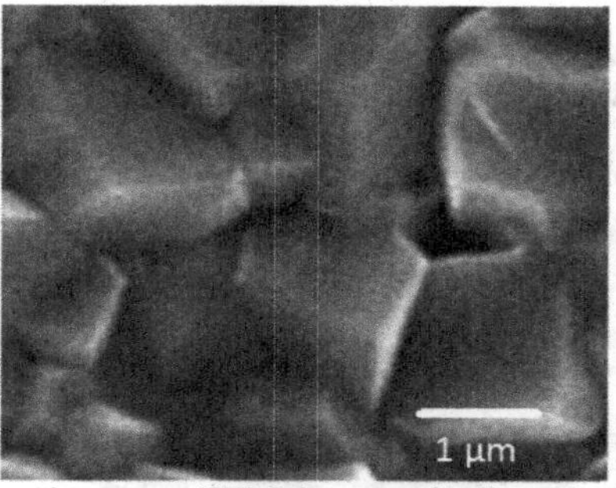


**Figure 1.** Scanning electron microscope (SEM) images of indium phosphide grown through metal organic chemical vapor deposition (MOCVD) in the form of a poly-crystalline layer grown on copper not treated with carbon.

While InP grew as thin films on copper substrate without a carbon layer, InP on copper substrates with a carbon layer grew as polycrystalline nanowires, as seen in Figure 2a-c. They appear to form from a central location, indicating that perhaps a single gold colloidal nanoparticle seeded multiple nanowires. EDS data indicates that their elemental composition is indium and phosphorus. Based on the atomic percentages given for the various elements, indium and phosphorus are approximately 1:1 indicating that we formed indium phosphide. Copper, gold, and oxygen were also present. Copper and gold indicate that we are measuring the surface of the substrate, and the substrate is made of copper and was treated with gold. Figure 2a indicates that a high density growth was achieved. The SEM image shown in Figure 2b would also indicate that the nanowires are precipitating out of the metallic colloids based on the change in shade of the tips of the nanowires. This nanowire points to the growth mechanism as some form of VLS method [7]. The image highlighted in Figure 2c shows the rough morphology of the growth.

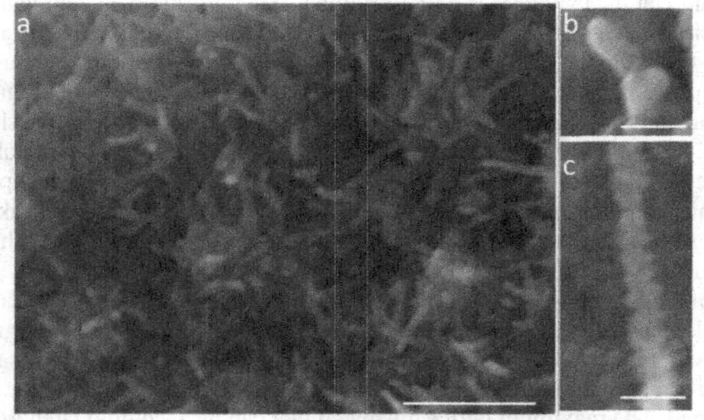


**Figure 2.** Scanning electron microscope (SEM) images of indium phosphide grown through metal organic chemical vapor deposition (MOCVD) a) in the form of poly-crystalline nanowires grown on copper treated with carbon, scale is 2.5 µm, b) poly-crystalline nanowire during the

beginning of growth, scale is .25 μm, and c) a mature poly-crystalline nanowire showing the unique morphology of this growth, scale is .25 μm.

Raman spectroscopy was performed to confirm that the carbon layer formed on the copper sheet was graphene. Figure 3 shows the raman data taken from a copper substrate after the treatment process described in the experimental section. This data shows raman peaks at the locations of the G and 2D vibrational states [8-10]. This would indicate that we have a highly disordered layer of graphene.

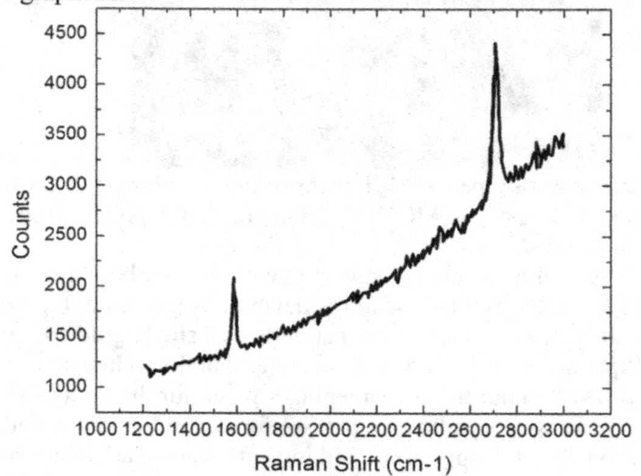


**Figure 3.** Raman spectroscopic data taken from the copper substrate treated with carbon showing two distinct graphene peaks.

TEM images of a representative InP nanowire are exhibited in Figure 4. Figure 4a depicts a lower magnification TEM image of a nanowire that was grown on copper. A higher magnification of the nanowire shown in 4a is shown in 4b. These two images show grain features that indicate that the nanowire is polycrystalline. They illustrate an aggregation of small grains with dimensions from ~2-11 nm into groups of larger grains with dimensions ranging from 40 to 120 nm that form the nanowires. This indicates a polycrystalline nature as compared with a granular structure in which plane alignment would be highly improbable. Figure 4c, a selective area diffraction (SAD) pattern of the nanowire shown in figures 4a and b, shows several rings indicating polycrystallinity which confirms our instinct that the grains are not all aligned to a single plane. The SAD also displays a feature that indicates the [-111] crystalline plane was preferred in its family of planes present in the area that the diffraction pattern was taken. Therefore we have some sort of "poly-preference" phase present in this nanowire. This could be due to some alignment with the seed from which the grains grew.

In growth of a thin film the limit of how much stress or strain can be accommodated by the growth material depends on this parameter and thickness. However with this large of a mismatch growth will not occur [11]. One advantage of nanowire growth is that it can accommodate significantly larger lattice mismatches as there are fewer atoms whose bonds will be stressed or strained. The maximum lattice mismatch that can be accommodated is still only 11.6%, but the mismatch is ~11% [12]. This is right on the edge of what can be accommodated, and therefore, the strain could have caused this granular formation. By adding carbon, we believe a layer has been created that allows for growth of small grains on disordered carbon, and this disorder could have also contributed to grain formation. It is also possible that the Au-Au bond interaction versus Au-C on graphene caused breakage as seen in Figure 2b [13]. Perhaps it is a combination of all three of these possibilities or another factor not yet accounted for.

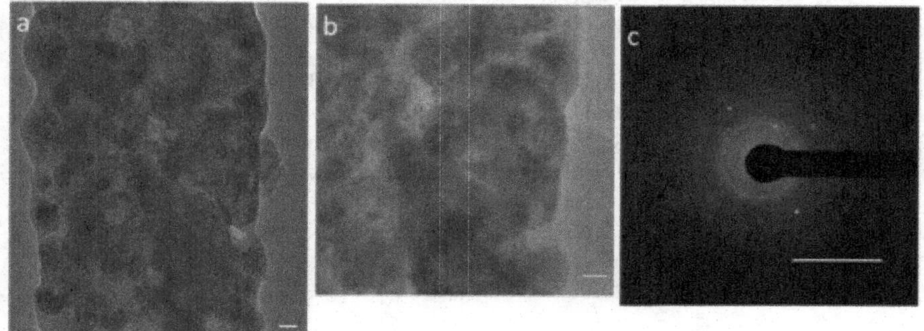


**Figure 4.** Transmission Electron Microscopy of a) a low magnification image of a polycrystalline indium phosphide nanowire showing grains ranging from ~20-40 nm, scale is 20 nm b) a high magnification image of the nanowire depicted in 4a that shows the amorphous region between grains, scale is 20 nm, and c) the diffraction pattern of this nanowire showing that it is polycrystalline. However, the bright spots on the rings indicate a particular plane, [-111], was more prevalent in the area studied here, scale is 1.0 nm Dif.

## CONCLUSIONS

Here we have presented data showing that a method to grow polycrystalline indium phosphide nanowires has been developed. These nanowires have grains of all sizes, and with further investigation of the growth mechanism, this could lead to control that would allow the grains to be engineered for optimum scattering of phonons. This technology to grow polycrystalline nanowires on copper could lead to a new device platform for thermoelectric energy generation.

## ACKNOWLEDGMENTS

This work was supported by NASA SBIR NNX11CE14P. We would like to thank HP labs and the MACS facility (Moffett Field, California) at Advanced Studies Laboratories, University of California Santa Cruz, and NASA Ames Research Center for continuous support on analytical equipment. This material is based upon work supported by the National Science Foundation Graduate Research Fellowship under Grant No. DGE-0809125. Support by Semiconductor Research Corporation CSR fund (Dr. Victor Zhirnov) is also highly appreciated.

## REFERENCES

1. Y. Li, G. W. Meng, L. D. Zhang and F. Phillipp. "Ordered semiconductor ZnO nanowire arrays and their photoluminescence properties." Applied Physics Letters **76**, 2011 (2000)

2. N. Wang, Y. H. Tang, Y. F. Zhang, C. S. Lee, I. Bello and S. T. Lee. "Si nanowires grown from silicon oxide." Chemical physics letters **299**, 237 (1999)
3. N. Wang, Y. H. Tang, Y. F. Zhang, C. S. Lee and S. T. Lee. "Nucleation and growth of Si nanowires from silicon oxide." Physical Review **B 58**, R16024 (1998)
4. K. Hillerich, M. E. Messing, L. R. Wallenberg, K. Deppert and K. A. Dick. "Epitaxial InP nanowire growth from Cu seed particles."Journal of Crystal Growth **315**, 134 (2011)
5. B. Poudel, Q. Hao, Y. Ma, Y. Lan, A. Minnich, B. Yu, X. Yan, D. Wang, A. Muto, D. Vashaee, X. Chen, J. Liu, M. S. Dresselhaus, G. Chen and Z. Ren. "High-thermoelectric performance of nanostructured bismuth antimony telluride bulk alloys." Science **320**, 634 (2008)
6. T. T. Ngo-Duc, J. Gacusan, N. P. Kobayashi, M. Sanghadasa, M. Meyyappan and M. M. Oye. "Controlled growth of vertical ZnO nanowires on copper substrate." Applied Physics Letters **102**, 083105 (2013)
7. R. S. Wagner and W. C. Ellis. "Vapor-liquid-solid mechanism of single crystal

growth." Applied Physics Letters **4**, 89 (1964)
8. Z. H. Ni, T. Yu, Y. H. Lu, Y. Y. Wang, Y. P. Feng and Z. X. Shen. "Uniaxial strain on graphene: Raman spectroscopy study and band-gap opening." Acs Nano **2**, 2301 (2008)
9. D. Graf, F. Molitor, K. Ensslin, C. Stampfer, A. Jungen, C. Hierold and L. Wirtz. "Spatially resolved Raman spectroscopy of single-and few-layer graphene." Nano letters **7**, 238 (2007)
10. M. S. Dresselhaus, A. Jorio, M. Hofmann, G. Dresselhaus and R. Saito. "Perspectives on carbon nanotubes and graphene Raman spectroscopy." Nano letters **10**, 715 (2010)
11. E. Kuphal. "Phase diagrams of InGaAsP, InGaAs and InP lattice-matched to (100) InP." Journal of crystal growth **67**, 441 (1984)
12. L. C. Chuang, M. Moewe, C. Chase, N.P. Kobayashi, C. Chang-Hasnain and S. Crankshaw." Critical diameter for III-V nanowires grown on lattice-mismatched substrates." Appl. Phys. Lett. **90**, 043115 (2007)
13. P. K. Mohseni, A. Behnam, J. D. Wood, C.D. English, J. W. Lyding, E. Pop and X. Li "InxGa1–xAs Nanowire Growth on Graphene: van der Waals Epitaxy Induced Phase Segregation." Nano letters **13**,1153 (2013)

Mater. Res. Soc. Symp. Proc. Vol. 1543 © 2013 Materials Research Society
DOI: 10.1557/opl.2013.949

# Quaternary Chalcogenide Nanocrystals: Synthesis of $Cu_2ZnSnSe_4$ by Solid State Reaction and their Thermoelectric Properties

Umme Farva and Chan Park
Department of Materials Science and Engineering,
Seoul National University,
Seoul 151-744, Republic of Korea

## ABSTRACT

In this paper, synthesis of $Cu_2ZnSnSe_4$ (CZTSe) materials by using simple and cost-effective solid state reaction method from the elemental Cu, ZnO, SnO and elemental Se powders are carried out. The SEM images show spherical, non-uniform size with aggregation of nanopowders. The phase separation and thermal analysis of the milled powders suggested that most of the starting powders reacted because of a mechanical alloying effect during milling process. After the solid state reaction at above 500 °C, the nanopowders crystallized into stannite single phase, which are confirm by XRD spectra. The thermoelectric properties of synthesized powder are under study.

## INTRODUCTION

The thermoelectric (TE) devices, allowing the solid-state conversion between thermal and electrical energy, have long been considered a very attractive technology for cooling and waste heat recovery [1, 2]. The ideal TE materials possess low thermal conductivity ($\kappa$) with good electrical conductivity ($\sigma$) and high Seebeck coefficient ($S$). Thermoelectric effects arise because charge carriers in metals and semiconductors are free to move much like gas molecules, while carrying charge as well as heat. The best thermoelectrics are semiconductors that are so heavily doped their transport properties resemble metals [3]; allow separating control of $\sigma$ (electron) and $\kappa$ (phonons). The chalcopyrite-type semiconductor compounds are known to have a potential for application in photovoltaic energy devices, light-emitting diodes and nonlinear optical devices as well as thermoelectric devices [4-7]. These ternary and quaternary chalcogenide compounds can be designed by various combinations of elements in the vicinity of the group IV element. Such tetrahedrally coordinated semiconductors, for example, defect chalcopyrite, spinel, stannite and famatinite, have been suggested [8, 9]. To examine the possibility of the above semiconductors for useful devices is very important for newer material developments. However, the physical properties of these semiconductor compounds, especially of many quaternary compounds, have been scarcely investigated, because it is quite difficult to obtain the crystals with suitable size and quality as compared with elemental and binary compounds [7, 10].

The Cu-doped quaternary chalcogenide are wide-band gap semiconductor belongs to stannite structure. These materials are p-type transparent conductive materials (TCMs) designed from the concept of two structural/functional units (electrically conducting and insulating). These materials are having excellent TE properties at medium temperatures owing to relatively high $S$ because of wide band gap and low $\kappa$ values due to their chalcopyrite structure. Here in we present a successful synthesis of $Cu_2ZnSnSe_4$ (CZTSe) nanopowders in stannite crystal phase using solid state reaction method.

## EXPERIMENT

The quaternary chalcogenide $Cu_2ZnSnSe_4$ nanopowders were prepared by solid state reaction process. We used elemental Cu and Se as the copper and selenium precursor and ZnO, $SnO_2$ for Zn and Sn source respectively. The stoichiometric ratios of Cu, ZnO, $SnO_2$ and Se with a purity of 99% were initially weighed and then the starting materials were transfer into a high density polyethylene container. The ball milling processes were carried out by using vertical mixer for 200h at 175 rpm milling speed in anhydrous ethanol with spherical zirconia balls (zirconia grinding media; YSZ). After completion of the process, final products were vacuum dried at 80 °C for overnight. Then a fixed amount (5g) of CZTSe nanopowders were placed in a silica crucible and loaded into a quartz tube furnace followed by heating from room temperature to various temperatures (300 °C, 400 °C, 500 °C, 600 °C and 700 °C, respectively) with a heating rate of 10 °C $min^{-1}$ in inert atmosphere for 1h to performed the solid state reaction and at 650 °C for 6h to compare the influence of time duration on the crystallinity of CZTSe nanopowder. After accomplished the process, the furnace was allowed to cool naturally. The crystal structural, surface morphology, elemental composition and thermoelectric properties of synthesized and thermally treated CZTSe powders were characterized by means of SEM, SEM-EDX, TEM, XRD measurements.

## DISCUSSION

Figure 1a shows the SEM image of as-synthesized CZTSe nanopowders after ball milling process. It is clearly revealed that using the milling process the particles are much smaller in size, platelet shape and also seen 1-2 micron size with fine nanofibers protruding from the pellets. The EDX measurements of these $Cu_2ZnSnSe_4$ nanopowders indicated the chemical composition ratios of the Cu:Zn:Sn:Se is nearly close to their stoichiometric ratio; 2:1:1:4, as depicted in Figure. 1b.

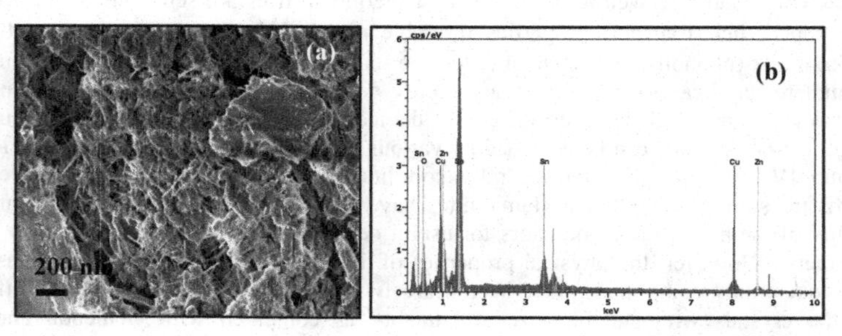


Figure 1. SEM (a) and EDX (b) characterization of as-synthesized $Cu_2ZnSnSe_4$ nanopowders.

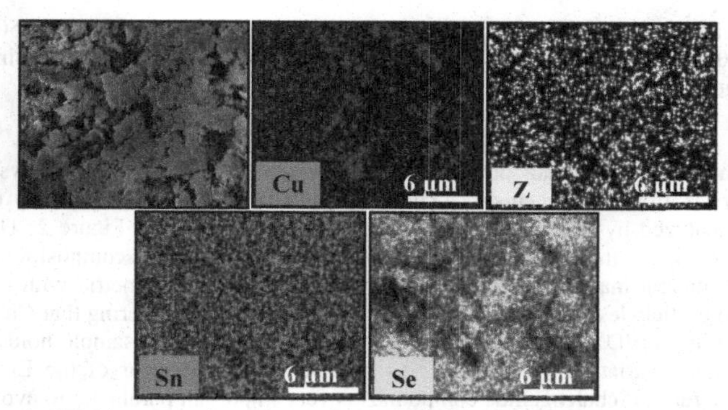

Figure 2. SEM image and corresponding EDX elemental mapping of Cu, Zn, Sn and Se.

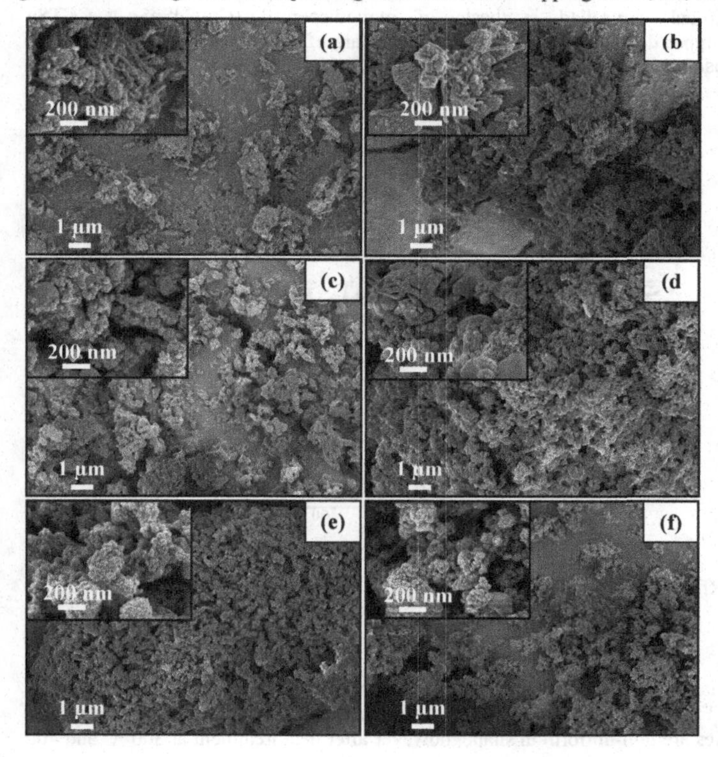

Figure 3. SEM images of annealed nanopowders (a) at 300 °C, (b) at 400 °C, (c) at 500 °C, (d) at 600 °C, (e) 700 °C for 1h respectively and (f) at 650 °C for 6h; Inset are corresponding high magnification images.

The oxygen is also present in nanopowders, which was confirmed by EDX analysis possibly this oxygen arises from the precursors (ZnO, $SnO_2$). The chemical composition of nanopowders was also analyzed by SEM elemental mapping which is shown in Figure 2. The elemental mapping clearly shows that the nanopowders are slightly Se-rich. The compositional analysis by EDX and elemental mapping are correlated approximately stoichiometric with the variation from particle to particle less than the experimental error of ca. 2, considering that Cu is slightly over represented in the EDX spectra because of the signal from the Cu sample holder. There was no compositional variation from particle to particle within the error of the EDX detector. The composition ratio of chalcogenide compounds is very important parameter to avoid copper rich or sulfur and/ or selenium deficiency of the nanoparticles [11]. Recently, Fan et al [7] reported first colloidal controlled synthesis of monodispersed ($Cu_2CdSnSe_4$) CCTSe nanocrystals, with copper dopants and selenium vacancies are effective in enhancing the power factor and the dimensionless figure-of-merit ZT value reach 0.65 at 450 °C.

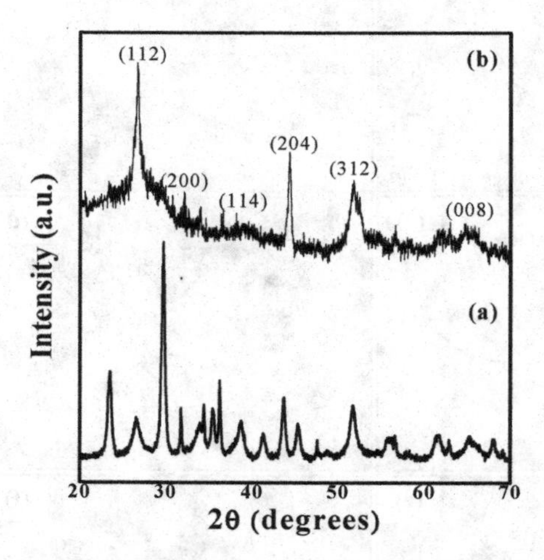


Figure 4. XRD patterns of (a) as-synthesized and (b) annealed at 500 °C for 1h CZTSe nanopowders

The SEM images of thermally-treated CZTSe nanopowder at different temperatures for 1h (300 °C-700 °C; Figure 3a-e) and 6h (650 °C; Figure 3f) are shown in Figure 3. It is clearly observed that particles are non-uniform in shape, however after heat treatment at 300 °C and 400 °C, the tripod shaped nanofibers (as-synthesized; Figure 1a) are transform to ~1-2 μm size nanorods that is simultaneously reduced with increase in heating temperature. Heating above 400 °C, the nanopowders are becomes more aggregated, dense, spherical in shape size with relatively uniform size of ~ 70 nm to 20 nm (500 °C-700 °C), which is also confirm by TEM (results are not shown). The shape transformation is occurring due to the high thermal effect. The annealed time duration is also affect on the surface morphology of CZTSe nanopowder, as shown in Figure 3f. It is interesting to observe that at the annealing temperature of 650 °C for 6h particles are more agglomerated, slightly bigger in grain size as compared with annealed at 600 °C or 700 °C for 1h.

X-ray diffraction patterns of as-synthesized CZTSe nanopowders and annealed at 500 °C temperature for 1h are compared in Figure 4. The XRD pattern of CZTSe nanopowders after ball milling process has shown the presence of Zn, Sn along with reaction between Cu and Se to form a klockmannite CuSe binary phase or α-CuSe [12]. As reported earlier [13] that the presence of the α-CuSe binary phase in the milled powders suggested that the powders underwent mechanical alloying during the ball milling process [14, 15]. It is observed that after annealed at high temperature the CZTSe nanopowders have crystalline stannite tetragonal structure with 2θ peaks at ~27, 45, 53.4, 38.8, 31.4, 65.8° corresponding to the (112), (204), (312), (114), (200) and (008) crystal planes, respectively [16]. The higher peak intensities of CZTSe nanopowders after annealing at 650 °C for 6h (result not shown) were attributed to the sintering effect, which can also consistent with the SEM and TEM results. The impurities elemental Zn, Sn and α-CuSe phase are removed after further treatment. The thermoelectric properties of prepared high crystalline stannite phase $Cu_2ZnSnSe_4$ nanopowders will be investigated and may increase *ZT* valve subsequently optimize several conditions.

## CONCLUSIONS

The nano-size $Cu_2ZnSnSe_4$ (CZTSe) materials have been successfully prepared from the elemental Cu, ZnO, SnO and elemental Se powders by using simple and cost-effective solid state reaction method. Annealing of the as synthesized milled nanopowders induces the shape and size of the nanopowders with improved crystallinity. It was observed by SEM and TEM, the particles are aggregated, spherical, non-uniform with ~70-20 nm in size after heat treatment above 500 °C and also crystallized into stannite single phase, which are confirm by XRD spectra. The shape and size of the nanostructures could be altered by annealing temperature and time duration, which plays an important role. Such useful information may help in the further study of CZTSe nanopowders for thermoelectric characterization.

## ACKNOWLEDGMENTS

This work was supported by BK21 Materials Education and Research Division Program at Seoul National University. The authors wish to express their thanks to Mr. Jun Young for their help in characterization of CZTSe materials.

## REFERENCES


1. Chung, D. Y.; Hogan, T.; Brazis, P.; Rocci-Lane, M.; Kannewurf, C.; Bastea, M.; Uher, C.; Kanatzidis, M. G. Science 2000, 287, 1024.
2. G. J. Snyder and E. S. Toberer, *Nat. Mater.* **7**, 105-114 (2008).
3. G. J. Snyder and E. S. Toberer, Nature Materials, 7, 105 – 114, 2008.
4. G. Zoppi, I. Forbes, R. W. Miles, P. J. Dale, J. J. Sragg and L. M. Peter, *Prog. Photovolt: Res. Appl.* **17**, 315–319 (2009).
5. S. F. Chichibu, T. Ohmori, N. Shibata, T. Koyama and T. Onuma, *J. Phys. Chem. Solids* **66**, 1868-1871 (2005).
6. T. Ouahrani, A. H. Reshak, R. Khenata, B. Amrani, M. Mebrouki, A. O-Roza and V. Luaña, *J. Solid State Chem.* **183**, 46-51 (2010).
7. F.-J. Fan, B. Yu, Y.-X. Wang, Y.-L. Zhu, X.-J. Liu, S.-H. Yu and Z. Ren, *J. Am. Chem. Soc.* **133**, 15910-15913 (2011).
8. H. Matsushita, T. Ichikawa and A. Katsui, *J. Mat. Sci.* **40**, 2003-2005 (2005).
9. M. L. Liu, I. W. Chen, F. Q. Huang and L. D. Chen, Adv. Mater. **21**, 3808-3812 (2009).
10. *Ternary Chalcopyrite Semiconductors: Growth, Electronic Properties, and Applications*, edited by J. L. Shay and J. H. Wernick (Pergamon, Great Britain, 1975), p. 13.
11. G. Zoppi, I. Forbes, R. W. Miles, P. J. Dale, J. J. Scragg and L. M. Peter, *Res. Appl.* **17**, 315-319 (2009).
12. The International Center for Diffraction Data [ICDD 00-034-0171].
13. R. A. Wibowo, W. H. Jung and K. H. Kim, *J. Phys. Chem. Solids* **71**, 1702-1706 (2010).
14. T. Ohtani, M. Motoki, K. Koh and K. Ohshima, *Mater. Res. Bull.* **30**, 1495-1504 (1995).
15. T. Ohtani, K. Maruyama and K. Ohshima, *Mater. Res. Bull.* **32**, 343-350 (1997).
16. The International Center for Diffraction Data [ICDD 04-003-8817].

**Mater. Res. Soc. Symp. Proc. Vol. 1543 © 2013 Materials Research Society**
**DOI: 10.1557/opl.2013.676**

# Design and Evaluation of Carbon Nanotube Based Nanofluids for Heat Transfer Applications

Sathya P. Singh[1], Nader Nikkam[1], Morteza Ghanbarpour[2], Muhammet S. Toprak[1*], M. Muhammed[1] and Rahmatollah Khodabandeh[2]
[1] Department of Materials and Nano Physics, KTH-Royal Institute of Technology, SE-16440 Kista, Stockholm, Sweden.
[2] Department of Energy Technology, KTH-Royal Institute of Technology, SE-100 44 Stockholm, Sweden.

## ABSTRACT

The present work investigates the fabrication, thermal conductivity (TC) and rheological properties of water based carbon nanotubes (CNTs) nanofluids (NFs) prepared using a two-step method. As-received (AR) CNTs heated and the effect of heat treatment was studied using X-ray diffraction and thermogravimetric analysis. The AR-CNTs and heat-treated CNTs (HT-CNTs) were dispersed with varying concentration of surface modifiers Gum Arabic (GA) and TritonX-100 (TX) respectively. It was found that heat treatment of CNTs effectively improved the TC and influenced rheological properties of NFs. Scanning electron microscopy analysis revealed TX modified NFs showed better dispersion ability compared to GA. Surface modification of the CNTs was confirmed by Fourier Transformation Infrared (FTIR) analysis. Zeta potential measurement showed the stability region for GA modified NFs in the pH range of 5-11, whereas pH was between 9.5-10 for TX NFs. The concentration of surface modifier plays an extensive role on both TC and rheological behavior of NFs. A maximum TC enhancement of 10% with increases in viscosity around 2% for TX based HT-CNTs NFs was measured. Finally comparison of experimental TC results with the predicted values obtained from a model demonstrated inadequacy of the predictive model for CNT NFs system.

## INTRODUCTION

Heat transfer technologies are being focused on the development of highly efficient and new hybrid materials for effective heat transfer or cooling system. Demand for thermal management in various industries such as microelectronics, combustion engines, boilers, power transformers, space and defense cannot be met by the conventional heat exchange fluids, like water and ethylene glycol due to their limited heat transfer performance. Nanofluids, as the term coined by Choi [1], in which a small amount of nanoparticles added into base fluids, showed promising heat transfer performance over conventional heat exchange fluids.

CNTs possess high electrical and thermal conductivity, which makes them an attractive candidate for NFs for heat transfer applications. CNTs cannot be dispersed in water directly due to their hydrophobic surface. Therefore, selecting suitable surface modifier with the optimized concentration plays an important role in fabrications of stable CNT based NFs.

To develop an effective CNTs NFs, we focused on the improvement of the quality of AR-CNTs by heat treatment. Then two different surface modifiers, Gum arabic (GA) and triton X-100 (TX), were used based on their binding ability with graphitic structures on the CNTs and stabilize them in water. TX is a surfactant having a methylene group, a graphitic liking structure, as a hydrophobic tail, in which a benzene ring adsorb CNTs surface more strongly using π-π stacking type interaction [2]. GA is a dispersant with a highly branched polysaccharide composed of mainly D-galactose, L-arabinose, L-rhamnose, and glucuronic acid. These complex long chains provide easy binding ability [3].Characterization of CNTs as well as the developed NFs are presented in detail.

## EXPERIMENTAL

Multi-walled CNTs, Gum Arabic (GA), a natural polysaccharide in the form of large granules (100% pure and organic) and Triton X-100 (TX, 100% pure, nonionic) were obtained from the Research Institute of Petroleum Industry, Iran.

A two-step method was adopted, which include sonication of CNTs (0.1 wt%) in DI-water, following addition of surface modifiers and further bath sonication for homogenization of the mixture. Finally, in order to achieve an optimum dispersion with thermodynamic stability, the pH was adjusted in the range of 9.5-10 by using 0.5 M NaOH.

HT-CNTs were prepared by heating the samples at 300 °C for 2 hours in air. To study the effect of heat treatment on CNTs, samples were characterized by X-ray diffraction (powder diffractometer Panalytical X'Pert PRO) with Cu Kα radiation (λ=1.5418 Å), scanned in the 2θ range of 10-90°. In order to study their thermal stability, thermogravimetric analysis (TA instrument, TGA Q 500) was carried out at 5 °C/min heating rate in air.

SEM analysis (Zeiss Ultra 55 SEM) was performed. Fourier transform infrared spectrometry (SMART iTR-Nicolet iS10), zeta potential analysis (Beckman Coulter DelsaNano C, thermal conductivity analysis (Transient Hot wire Method -THW), rheological measurements (Brookfield model DV-II + Pro with UL adapter) were performed to investigate morphology, dispersion characteristics, thermal conductivity and rheology of the CNTs dispersion.

## RESULTS AND DISCUSSION

Heat treatment of CNTs at 300 °C in air has been reported to improve the average specific strength and the symmetry of graphitic sheet when compared to as-synthesized CNTs improving overall quality of CNTs [4]. Thermogravimetric analysis on AR-CNTs (inset in Figure 1(A)) shows the effect of heating on CNTs. Up to 300 °C approximately 1% weight loss was observed. Further increase in temperature above 450 °C brings about an abrupt change in the weight loss, indicating oxidation of carbon and formation of gaseous products. Therefore a calcination temperature of 300 °C was selected. Figure 1(A), displays the X-ray diffraction patterns for both AR-CNTs and HT-CNTs. The patterns are very similar with slight improvement in the broadness and intensity of the characteristics graphitic peaks, mainly MWCNTs at 26.3° & 43.6° [5]. The resemblance of this peaks pattern and temperature stability shows strong evidence that there is no change in the material structure; the HT-CNTs retain their graphitic structure even after heat treatment.

### FTIR Analysis

Stable dispersion of CNTs NFs may lead to achieve improved thermal and rheological properties of the NFs. Surface modifiers minimize the Van der Waals interaction forces between CNTs, which can enhance CNTs dispersability in water as the base-liquid.

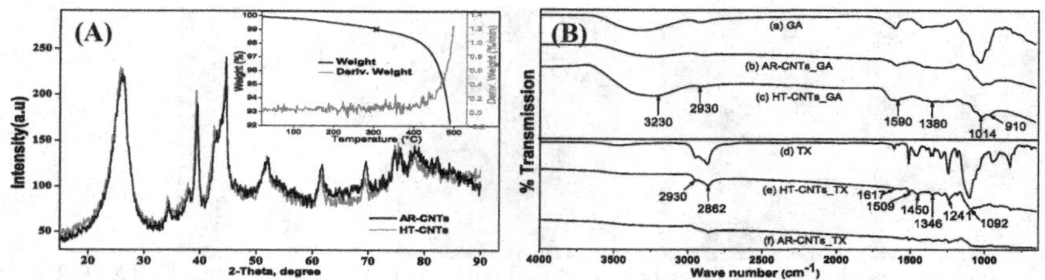

**Figure 1.** (A) XRD pattern of AR-CNTs and HT-CNTs: Inset: TGA of CNTs. (B) FTIR spectra of (a) GA, (b)AR-CNTs_GA, (c) HT-CNTs_GA, (d) TX, (e) HT-CNTs_TX, (f) AR-CNTs_TX.

In order to assure the attachment of the surface modifiers to CNTs surface we performed FTIR analysis and the results are shown in figure 1(B). The spectra of HT-CNTs and AR-CNTs are very similar, except for the peak intensities. FTIR spectra for GA modified CNTs are presented in figure 1(B)(b) and 1(B)(c), where the following assignments are made for the observed absorption peaks: 3230 $cm^{-1}$ -OH stretching vibration from the alcohols in polysaccharides and N-H group from the amines; 1590 $cm^{-1}$ -C=O stretching vibration, 1380 $cm^{-1}$ -$CH_3$ stretching vibration due to rhamnose of the monosaccharide [6]; the strong peak at 1014 $cm^{-1}$ -C-OH stretching vibration of the alcoholic group and the deformation vibration of C-O and C-C from the pyranose ring; 2930 $cm^{-1}$ -C-H aromatic stretching [6].

In TX modification, the hydrophobic octyl part of the surface modifier adsorbed CNTs by wrapping around the surface and hydration of oxyethylene chains favors hydrophilic behavior [7], as revealed by a large number of peaks involved in bonding HT-CNTs (e) and AR-CNTs (f). The presence of higher intensities of oxygenated functional groups at 1617 $cm^{-1}$, 1241 $cm^{-1}$ and 1092 $cm^{-1}$ are ascribed to C=O and C-O stretching vibration. The peak at 1509 $cm^{-1}$ is due to ring C=C stretching vibration in benzene ring structure, while 1450 $cm^{-1}$ is due to aromatic =CH stretching vibration and 1346 $cm^{-1}$ is from -C-H bending. The two peaks at 2915 $cm^{-1}$ and 2862 $cm^{-1}$ are due to $CH_3$ and $CH_2$ stretching vibration appearing in both the spectra, presence of oxyethylene chains [7]. In summary, the peaks related to GA (Figure 1(B)(a)) and TX (Figure 1(B)(d)) are dominant in the spectra of AR-CNTs and HT-CNTs respectively, revealing the success of the surface modification process.

In order to study the dispersion of CNTs equal concentrations of GA and TX were used for surface modification. CNTs were then dried and analyzed by SEM to study their dispersion/ aggregation behavior to understand the influence of different dispersants. SEM images of HT-CNTs modified by with GA and TX are shown in figure 2(A) and 2(B) respectively. GA modified HT-CNTs seem to be strongly agglomerated, as only few CNTs are visible clearly. TX modified HT-CNTs seem to be well separated, where individual fibers can be seen. Inset images shows clearly entangled individual CNTs. The average diameter of the CNTs is in the range of 10-45 nm, as observed from SEM images. The difference between the TX and GA dispersability is most likely due to the different molecular structures.

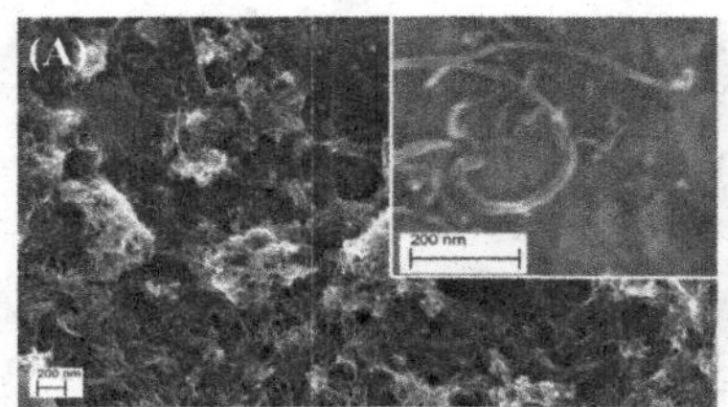

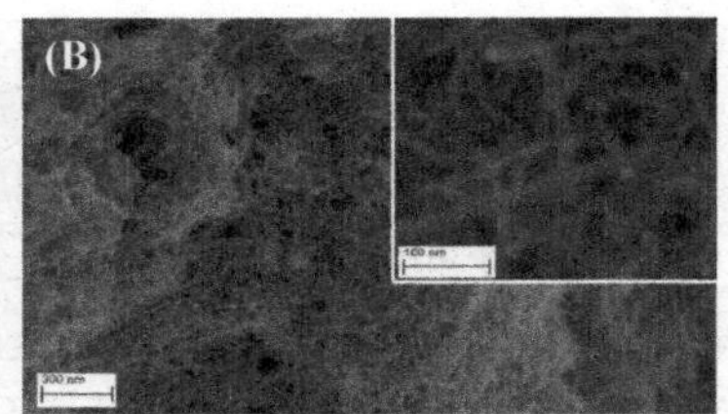


**Figure 2.** SEM images of dried CNTs surface modified with (A) GA (0.2 wt%) and (B) TX (0.2 wt%).

In order to study the stability conditions for CNTs NFs, zeta potential analysis was performed (data not shown) on diluted NF samples. The colloidal stability range for HT-CNTs and AR-CNTs with GA was obtained in the pH range of 6-10, while for TX modified HT-CNTs and AR-CNTs NFs, the maximum surface charge was observed at pH 10.

## Thermal Conductivity (TC) Measurements of NFs

Thermal conductivity, TC, measurements have been performed on a series NFs, to study the effect of type/treatment of CNTs on the TC of NFs. Figure 3(a) and 3(b) shows the ratio of TC of the NFs ($TC_{nf}$) with respect to the base liquid ($TC_{bl}$). NFs containing HT-CNTs modified with GA and TX showed higher ($TC_{nf}$)/($TC_{bl}$) values, indicating improvement over the base liquid. NF containing TX(0.05 wt%) modified AR-CNTs showed improvement above 30 °C, providing a sufficient evidence that heat treatment of CNTs has improved the TC of the NFs. Change in the concentration of surface modifiers notably influences the TC. In GA modified NFs, the maximum TC enhancement of 2.75% and 0.25% was observed with the concentration of 0.05 wt% GA at 40 °C over its base liquid and DI-water.

Similarly in the case of TX modified HT-CNTs NFs, maximum TC enhancement of 10.11% and 5% are obtained at 0.05 wt% TX at 40 °C, over its base liquid & DI-water (figure 3(b)).

It should be noted that NFs containing HT-CNTs has shown an improvement of 2% and 4.8% after modification with GA and TX over AR-CNTs, indicating that heat treatment of CNTs could provide improvement in TC of the CNT NFs.

## Predictive model for TC of CNT NFs

Xue *et al.* [8] developed a model which investigates the dependence of TC on CNT's length, diameter, concentration and interface thermal resistance simultaneously, given by equation (1), for the effective TC ( $K_e$ ) of the CNTs composites.

$$9(1-f)\frac{K_e - K_m}{2K_e - K_m} + f\left[\frac{K_e - K^c{}_{33}}{K_e + 0.14\frac{d}{L}(K^c{}_{33} - K_e)} + 4\frac{K_e - K^c{}_{11}}{2K_e + 0.5\left(K^c{}_{11} - K_e\right)}\right] = 0 \quad (1)$$

where $K^c{}_{11} = \frac{K_c}{1 + \left(2 * R_k * K_c / d\right)}, K^c{}_{33} = \frac{K_c}{1 + \left(2 * R_k * K_c / L\right)};$

Based on the given parameters, the TC of individual CNTs ($K_c$) = 3000 $Wm^{-1}K^{-1}$, Kapitza resistance ($R_k$) = $5.10^{-8}$ $m^2KW^{-1}$, TC of matrix composite ($K_m$) = 0.5 $Wm^{-1}K^{-1}$, length of the CNTs (*L*) = 50 μm and diameter of CNTs (*d*) = 45 nm. Predicted results are displayed in figure 4(a) revealing a linear slope for enhancement in TC ($K_e / K_m$) with increase in volume fraction (f) of CNTs/water suspension.

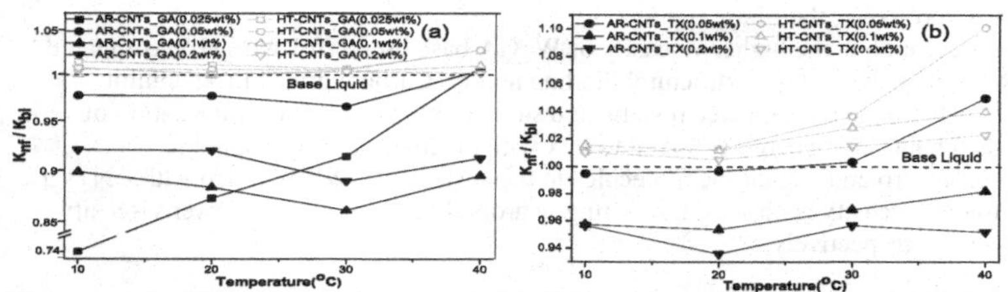


**Figure 3.** $TC_{nf}/TC_{bl}$ measurement on (a) GA (0.025-0.2 wt%) modified and (b) TX (0.05-0.2 wt %) modified AR- and HT-CNTs (0.1 wt%). (The lines are intended as guide to eyes).

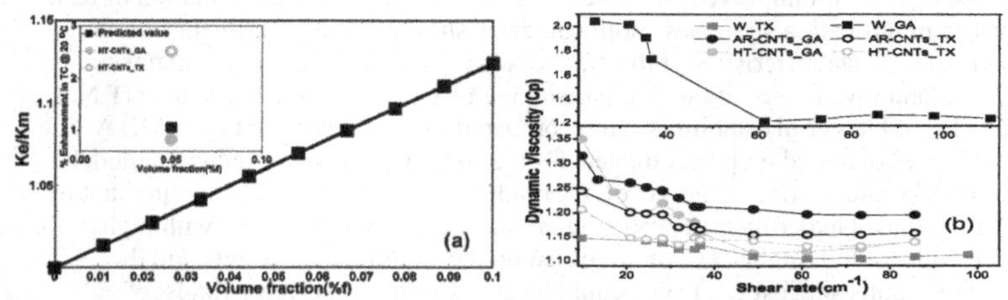


**Figure 4.** (a) Predicted enhancement in TC ($K_e / K_m$), using the model in Eq. (1) as a function of CNTs volume fraction (%f) at 20 °C; Inset: Comparison between the experimental results and predicted value for the TC of the CNTs NFs (b) Dynamic viscosity as a function of shear rate for AR- and HT-CNTs (0.1 wt%) modified with GA and TX.

A comparison between experimental results of the enhancement of TC of CNTs NFs and predicted value obtained by the model (assuming CNTs density = 2.1 $g/cm^3$) is presented in the inset of Figure 4(a). The TC enhancement value predicted by model is about 1.2% for 0.05% volume fraction for prepared CNT NFs. Measured TC for NFs containing HT-CNTs modified with TX (0.5 wt%) shows 2.5% enhancement, which is higher than the estimated value by a factor of 2, while HT-CNTs modified with GA (0.05 wt%) shows a 0.8% enhancement which is less than the predicted data by the model. It can be seen that model values under/overestimate the TC of CNTs NFs. There are many more influential properties such as interface, agglomeration that need to be included in TC enhancement predictive models for NFs.

## Rheological Measurements

Surface modifiers, plays a key role in stability, dispersion, TC and rheological properties. Therefore, we performed measurements on the base fluids containing the dispersant as well as the final NFs; the data are shown in figure 4(b). A common feature observed is the non-Newtonian characteristics of these fluids, which exhibit a high viscosity at low shear rate that decrease at high shear rates. Figure 4(b), shows that W_TX base liquid has the minimum viscosity, when compared to NFs with TX modified CNTs. HT-CNTs show minimum increase in viscosity with respect to their base liquid. The increase in viscosity of NFs with GA and TX modified HT-CNTs was observed to be 4.8% and 2% higher than their base liquids, respectively. A similar behavior was observed for NFs with GA and TX modified AR-CNTs, which show increase of 9.5% and 6% over their base liquid.

A peculiar observation is the higher viscosity of W_GA base liquid (0.1 wt% GA) compared to all NFs. This could be due to GA's structural binding ability. Carbohydrate blocks inhibit flocculation and coalescence through electrostatic and steric repulsions, favoring formation of individual strands (chain), but when it is forced to mix with other surface, hydrophobic polypeptide chains adsorb and anchor the molecules to the surface [3]. When mixed with CNTs a dramatic reduction in viscosity is observed, resulting in around 4.5% and 4.03% lower viscosity in HT- and AR-CNTs, respectively.

## CONCLUSIONS

Stable CNT based NFs with improved TC have been formulated using a combination of heat treatment and surface modification strategies. Both strategies showed positive influence on the stability, dispersion and TC characteristics of the NFs. CNTs NFs with TX exhibited better dispersion, higher TC and lower viscosity in comparison with CNTs NFs with GA, but CNT NFs modified with GA showed larger pH stability region compared to CNTs modified with TX. A TC enhancement of 10% was obtained with heat treated TX modified CNTs NFs. Predictive model developed for CNTs NFs underestimate the TC of TX modified NFs while overestimating that of GA modified CNTs obtained experimentally. NFs based on CNTs can be prepared with higher loading that in turn will result in higher TC values based on the experimental results and the predictive model. The results suggest CNT NFs could be effective heat exchange fluids.

## ACKNOWLEDGMENTS

This work is supported by the European Commission FP7 program under NanoHex project. The authors thank Prof. A. Rashidi from RIPI, Iran, for providing CNTs used in this work.

## REFERENCES

1. S.U.S. Choi and J.A. Eastman, American Society of Mechanical Engineers, 231, 99 (1995).
2. R.J. Robson and E.A. Dennls, Journal of Physical Chemistry, 81(11), 1075 (1977).
3. P.A. Williams and G.O. Phillips , Handbook of hydrocolloids, edited by G. O Phillips and Williams, P. A. (Woodhead Publishing Ltd, 2009) pp. 252.
4. Z. Yang, X. Sun, X. Chen, Z. Yong, G. Xu, R. He, Z. An, Q. Li and H. Peng, Journal of Materials Chemistry, 21, 13772 (2011)
5. E. Iyyamperumal, S. Wang, and L. Dai, ACS Nano, 6 (6), 5259 (2012).
6. C. Silvia, PhD. Thesis, University of Turin, 2006.
7. Y. Geng, M.Y. Liu, J. Li, X.M. Shi, J.K. Kim, Composites: Part A, 39, 1876 (2008).
8. Q.Z. Xue, Nanotechnology, 17, 1655 (2006).

**Nanoscale Transport Phenomena**

Mater. Res. Soc. Symp. Proc. Vol. 1543 © 2013 Materials Research Society
DOI: 10.1557/opl.2013.677

# Solute Effects on Interfacial Thermal Conductance

Andrew J. Green[1] and Hugh H. Richardson[1]
[1]Ohio University, Department of Chemistry and Biochemistry,
Athens, OH 45701, U.S.A.

## ABSTRACT

The thermal conductance of a gold/water interface has been found to change as a function of the surrounding's adhesion energy. We measure the thermal conductance of a lithographically prepared gold nanowire with a thin film nanoscale thermal sensor composed of AlGaN:$Er^{3+}$. The temperature of the nanowire is measured as a function of incident laser intensity. The slope of this plot is inversely proportional to the thermal conductance of the nanoparticle/surrounding's interface. We show that the conductance of the nanoparticle/water interface increases with the molality of the solution. This was tested with multiple solutes including NaCl, and D-Glucose. The interfacial conductance of pure water is reported to be 44 MW/$m^2$K and the conductance saturates to 100 MW/$m^2$K at a molality of 0.21 *m*.

## INTRODUCTION

Gold nanoparticles have been of great interest in the past decade due to their applications in photothermal therapy, drug delivery, and vapor generation. The ability to perform well in all of these applications will be directly related to the ability of the nanoparticle to release heat into the surroundings. Recent publications have shown that the interfacial conductance should depend on the surroundings adhesion energy at that interface.[1] Cahill et. al. have recently measured a change in thermal conductance across an SAM/Au interface as a function of the hydrophobicity of the surroundings.[2] This result inspires our study of solute contribution in interfacial thermal conductance.

This experiment will examine the heat transfer properties of a gold nanoparticle as the surroundings are changed from a pure water solution to different concentrations of NaCl, and D-Glucose. The different concentrations of solute significantly change the heat transfer of the gold nanowire thus altering the thermal conductance.

## EXPERIMENT

Gold nanowires with dimensions of 108 nm x 460 nm x 60 nm were fabricated with conventional e-beam lithography. A4 PMMA 950K was spin coated onto the substrate to a 100nm thickness. The substrate was then placed into an SEM and a pattern was drilled into the PMMA with a 50 keV acceleration voltage. The PMMA was developed with 3:1 IPA:MIBK for 3 minutes with a 30 second IPA rinse. An electron evaporation tool was used to sputter a thin adhesion layer of titanium (5 nm) along with 60 nm of gold through the mask. The residual PMMA layer was removed with a hot acetone bath (60 °C). This resulted in pristine gold nanowires as can be seen below in figure 1.

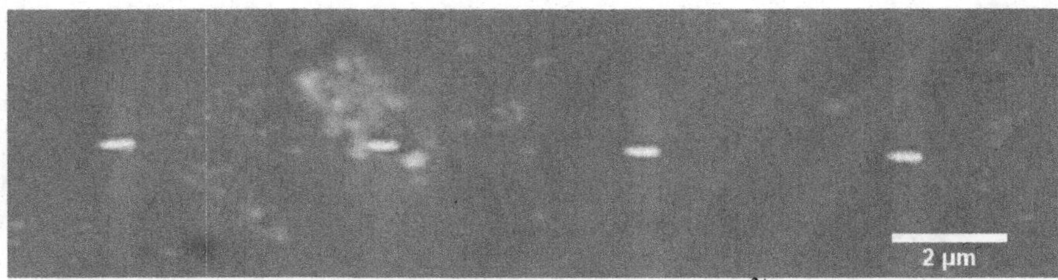

Figure 1: SEM image of a gold nanowire on a thin film of AlGaN:$Er^{3+}$ on silicon. The dimensions of the nanowire are 108 nm x 460 nm x 40 nm.

A temperature sensing thin film of AlGaN:$Er^{3+}$ is used to measure the temperature of the lithographically prepared gold nanowire. $Er^{3+}$ has two excited states that can be accessed with a 532 nm laser. The excited states are thermally coupled so the relative photoemission is directly related to the absolute temperature through a Boltzmann factor. Figure 2 below shows a spectral image collected with a scanning near field optical microscope (SNOM) upgraded with Raman Imaging.

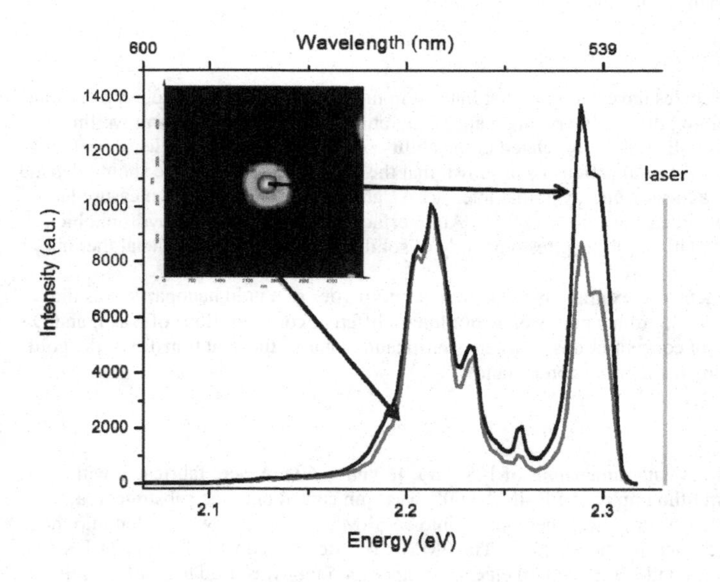


Figure 2: The inset in the upper left hand corner is a spectral image collected with the SNOM. Each pixel in the image contains a spectrum. The image shows a gold nanoparticle at the center.

A 'cold' pixel is shown by the red spectra and the two peaks are approximately equal in intensity. The 'hot' pixel in the black spectra shows an enhancement in the higher energy peak due to the thermalization of the excited states. This process has been used in previous publications, for more information please see references 3,4,5, and 6.

The spectral image is converted into a temperature image using a Boltzmann factor. The temperature of the nanowire is measured as a function of excitation laser intensity. Figure 3 below shows the change in temperature as a function of laser intensity with the particle in different environments.

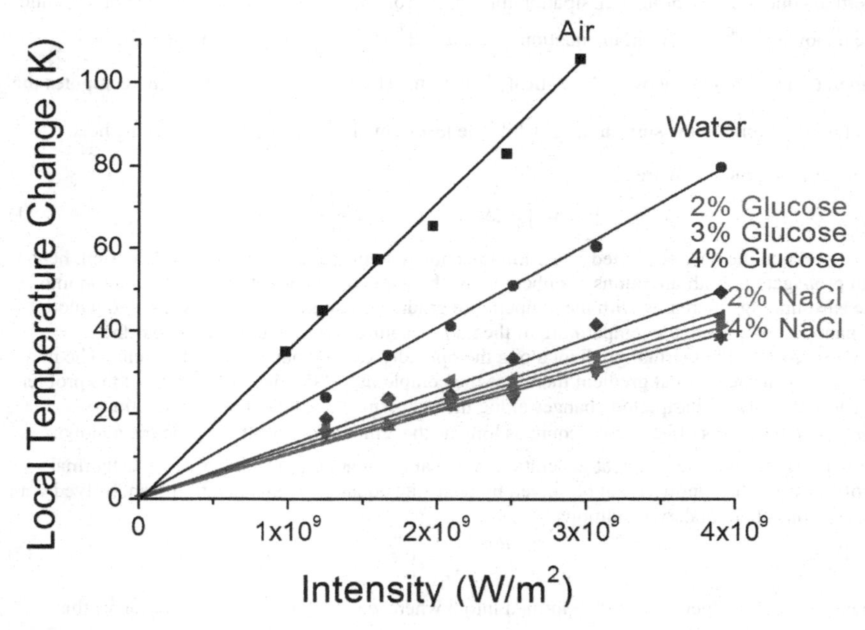


Figure 3: Local temperature changes of gold nanostructures as a function of laser excitation intensity. The black line shows data that was collected when the substrate was surrounded by air, and the blue line shows data when the substrate was submerged in water. Solutions with different concentrations of D-Glucose were measured (red) along with different concentrations of NaCl (blue). The concentrations vary from 2% to 4% by mass.

It should be noted that when the substrate is submerged in water, the slope is deflected down. This means at the same excitation intensity, the particle in water is at a lower temperature. When the particle is surrounded by water, the ability to transfer heat go up drastically. This creates a lower steady state temperature of the particle because a lower driving force is needed to remove the same amount of heat with respect to the gold particle surrounded by air. A direct comparison of these two examples is slightly unfair because the absorption cross sections are different. This will be addressed below in the discussion

**DISCUSSION**

As noted earlier, the temperature of the gold nanowire is going to be directly related to the heat transfer into the surroundings. To pull the thermal conductance out of the temperature measurement, rudimentary heat transfer equations will be used below. The temperature profile along the wire can be modeled using equation 1 where x is the direction along the wire, $k_{wire}$ is the thermal conductivity of the nanowire, A is the cross sectional area of the wire, g is the thermal conductance for heat dissipation into the surrounding, and P is the perimeter at x around the nanowire. The first term in equation 1, $\int \dot{q} A dx' = \int \frac{C_{abs} I(x')}{V_{exc}} A dx' = C_{abs} I_{avg}$, models the heat generation within the nanowire by optical excitation. The second term, $\int gP(\Delta T)dx$, computes the heat dissipation into the surroundings while the last term, $\int -k_{wire} A\left(\frac{d^2T}{dx^2}\right)dx$, relates the heat transfer along the nanowire.

$$\int \dot{q} A dx + \int gP(\Delta T)dx = \int -k_{wire} A\left(\frac{d^2T}{dx^2}\right)dx \qquad (1)$$

When the wire is excited in the middle and the temperature measured at that point, heat can propagate in both directions to either end of the wire. The amount of heat dissipation into the surroundings increases with the temperature gradient. Unlike the optically-excited nanodot or nanoparticle where the temperature of the nanostructure is nearly uniform across the nanostructure[13], the thermal gradient along the optically-excited nanowire is different. The non-uniformity of the thermal gradient increases the complexity of the dissipation part of the problem because the rate of dissipation changes along the nanowire. The heat dissipation term $\int gP(\Delta T)dx$ takes this effect into account as long as the temperature profile within the nanowire is known. At steady state, the heat generation and heat dissipation terms are equal. The thermal profile within the nanowire can be solved by recasting equation 1 into equation 2 and solved with the appropriate boundary conditions.

$$\frac{d^2\theta}{dx^2} + \frac{gP}{k_{wire} A}\theta = 0 \qquad (2)$$

In equation 2, temperature is transformed into θ where $\theta = T - T_\infty + \frac{\dot{q}}{k_{wire}}$. The solution of this differential equation is $\theta = C_1 \exp(-mx) + C_2 \exp(mx)$ where $m^2 = \frac{gP}{k_{wire} A}$. Applying appropriate boundary conditions yields the solution $\frac{\theta}{\theta_o} = \frac{\cosh(m(L-x))}{\cosh(mL)}$ where $\theta_o$ is $\theta$ at x=0 and L is the length of the nanowire.

In the heat generation term the laser intensity is not constant over the gold wire but varies as a Gaussian profile along the wire. The variable x' refers to the distance along the wire that is excited with the laser light and $I_{avg}$ is the average constant laser intensity illuminating the nanostructure equivalent to the spatial size where half of the laser intensity excites the nanowire.

The average laser intensity is determined by integrating the laser intensity over the dimensions of the nanowire and then restricting the dimension along the nanowire until half of the integrated intensity is obtained. The laser intensity at this restricted dimension is $I_{avg}$. The absorption cross section is $C_{abs} = \frac{12.8gP\int(\Delta T(x))dx}{I_{avg}}$. The factor 12.8 in the heat dissipation term relates the local temperature of the nanowire to the measured temperature.[13] This factor takes into account that our optical measurement of temperature is resolution limited and needs to be convoluted with the collection volume of our microscope and the true thermal image in the substrate.

The absorption cross section is calculated using energy balance after fitting the temperature profile. The temperature in the middle of the nanostructure is used to determine the absorption cross section using $C_{abs} = \frac{12.8g_s P_s\int(\Delta T(x))dx}{I_{avg}}$ where $P_s$ is the perimeter of the nanowire at location x in contact with the substrate. The integral $\int(\Delta T(x))dx$ is solved after the parameter $m^2$ is determined from the fitting. The absorption cross section at 532 nm for different length nanowires is nearly uniform ($6 \pm 1.4 \times 10^{-14}$ $m^2$) because only a minor portion of the nanowire is excited optically. The absorption cross for the nanowires submerged in water compared to air is expected to increase by a factor of 2 due to an increase in the effective dielectric constant of water compared to air.[21] The absorption cross section of the nanowires in water is $1.2 \times 10^{-13}$ $m^2$.

The thermal conductance of the nanowire with dissipation both into the substrate and the surrounding, g, can be solved once the thermal conductance for heat dissipation into the substrate is known. The heat dissipation can be partition into two parts.

Figure 4 shows the measured slopes as a function of solute molality. The slopes are taken from figure 3. Notice that the slope at 0 molality is just the pure water case. The slopes are the 0 point are 2%, 3%, and 4% D-Glucose by weight followed by 2%, and 4% NaCl by weight.

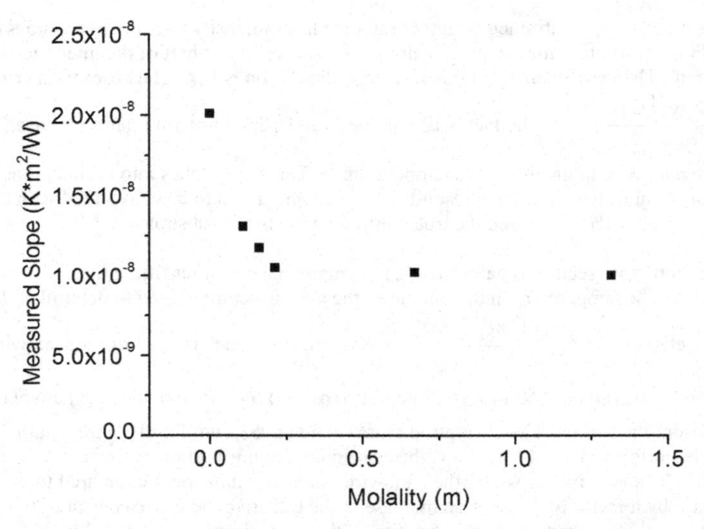


Figure 4: A plot of the measured slope (temperature as a function of excitation energy) vs molality of the solute in water. From left to right, the measured slopes from water, 2%,3%, and 4% D-Glucose, and 2%, and 4% NaCl

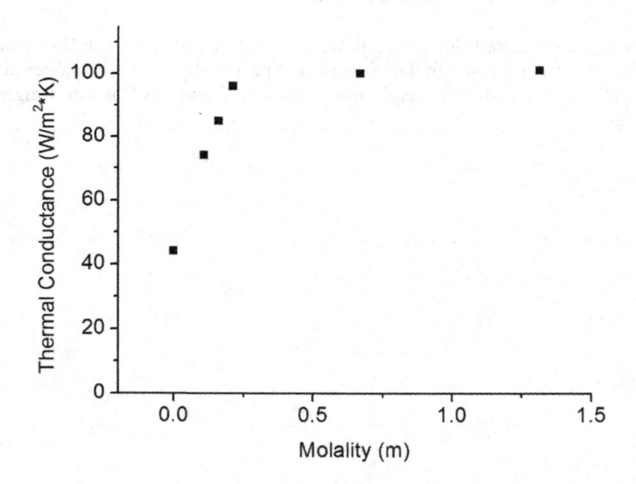

Figure 5: The calculations above allow for the transformation of the measured slope into a conductance for the interface. This plot shows the thermal conductance of that interface as a function of solute molality.

Figure 5 above shows the conductance of the gold/water interface as a function of solute molality. The conductance of the water interface is $44 MW/m^2K$. As solute is added to the solution the conductance saturates to a maximum value of $100\ MW/m^2K$. This value is in agreement with what is expected from a hydrophilic Au/Water interface.

## CONCLUSIONS

Gold nanostructures were fabricated with conventional ebeam lithography onto a temperature sensing thing film composed of $AlGaN:Er^{3+}$. The temperature of the nanowire was measured as a function of incident laser intensity. The slope of that line is direcly related to the conductance of the nanowire. Using general heat transfer equations we calculate an increase of the thermal conductance to go from $44\ MW/m^2K$ to $100\ MW/m^2K$.

## ACKNOWLEDGMENTS

The authors thank OSU center of Nanofabrication for their help in fabrication of the nanowires and the Condensed Matter and Surface Science Program at Ohio University.

## REFERENCES

1. L. Hu, L. Zhang, M. Hu, J-S Wang, B. Li, and P. Keblinski, Phys. Rev. B **2010,** 81, 235427
2. Losego, Grady, Sottos, Cahill. *Nature Materials.* **2012**
3. H. H. Richardson and D. Wang, Journal of Molecular Structure **2010**
4. M. T. Carlson, A. Khan, and H. H. Richardson, Nano Letters **2011**
5. Carlson, M.T.; Green, A.J.; Richardson, H.H.; Aurangzeb, K.;. *J. Phys. Chem. C.* **2012**
6. Carlson, M.T.; Green, A.J.; Richardson, H.H. *Nano Letters.* **2012**

Mater. Res. Soc. Symp. Proc. Vol. 1543 © 2013 Materials Research Society
DOI: 10.1557/opl.2013.674

# Heat transfer between a hot AFM tip and a cold sample: impact of the air pressure

Pierre-Olivier Chapuis[1,2], Emmanuel Rousseau[1,3], Ali Assy[2], Séverine Gomès[2], Stéphane Lefèvre[2], and Sebastian Volz[1]
[1]Laboratoire EM2C, Ecole Centrale Paris-CNRS, Grande voie des vignes, 92295 Chatenay-Malabry, France
[2]Centre de Thermique de Lyon (CETHIL), CNRS-INSA de Lyon-UCBL, 9 rue de la Physique, Campus La Doua-LyonTech, 69621 Villeurbanne (Lyon), France
[3]Groupe d'Etude des Semiconducteurs - CC074, Université de Montpellier II, Place Eugène Bataillon, 34095 Montpellier, France

## ABSTRACT

We observe the heat flux exchanged by the hot tip of a scanning thermal microscope, which is an instrument based on the atomic force microscope. We first vary the pressure in order to analyze the impact on the hot tip temperature. Then the distance between the tip and a cold sample is varied down to few nanometers, in order to reach the ballistic regime. We observe the cooling of the tip due to the tip-sample heat flux and compare it to the current models in the literature.

## INTRODUCTION

At nanometer scale probing the temperature [1] or measuring thermal properties such as the thermal conductivity [2] is a difficult task that may be tackled with scanning thermal microscopy (SThM) [3-4]. This technique is based on an atomic force microscope (AFM) with a tip sensitive to heat fluxes and/or temperature variations. SThM has been proposed for data storage [5–7], nano-lithography [8–10] or chemical sensing [11]. A key issue in local thermal analysis is to know the thermal sensitivity and the ultimate spatial resolution of the apparatus [12–14]. The resolution depends on the relative intensities of the heat fluxes transferred from the tip to the sample by various channels: conduction at their solid-solid contact; conduction in the liquid meniscus due to ambient humidity and located around the mechanical contact; air conduction around the tip [14]. The spatial resolution is linked to the contact area and improves when conduction though air and the water meniscus are removed. Working in vacuum (pressure lower than $10^{-2}$ mbar) removes the contribution from the surrounding air and might affect the water meniscus. It improves spatial resolution [15-16] but decreases the thermal sensitivity. An intermediate vacuum can be a compromise between spatial resolution and thermal properties sensitivity. In this study we focus on the heat flux due to conduction in air. We analyze the tip losses to the ambient environment and observe the cooling of the tip when approaching it close to the surface.

## EXPERIMENTAL SETUP

We now turn to the description of the first experiment. Our setup consists in a SThM embedded in a vacuum chamber. The probe is a 75 μm diameter Wollaston wire which is etched at its ends on approximately 200 μm. The outer part (silver) has been removed so that the inner

Pt90/Rh10 part of diameter 5 µm is apparent and bent in a V shape at its extremity. This probe has been extensively described in the past [17-18]: An electrical current heats the etched part of the wire though Joule effect. The probe can be heated alternatively by a DC or an AC current [18] and acts in the same time as the heater and the temperature sensor. When the average temperature of the probe varies a change in the probe electric resistance is induced, which leads to a modification of its voltage. Thus we are able to monitor the tip average temperature variations by monitoring the probe voltage. We use a Wheatstone bridge balanced at low current (no heating: Tip at room temperature) to observe a voltage signal directly proportional to the increase of temperature. When the tip is in contact with a sample, the probe cools down because heat is flowing through this contact. This allows observing thermal properties of samples.

**RESULTS**

**Heat losses in the ambient air**

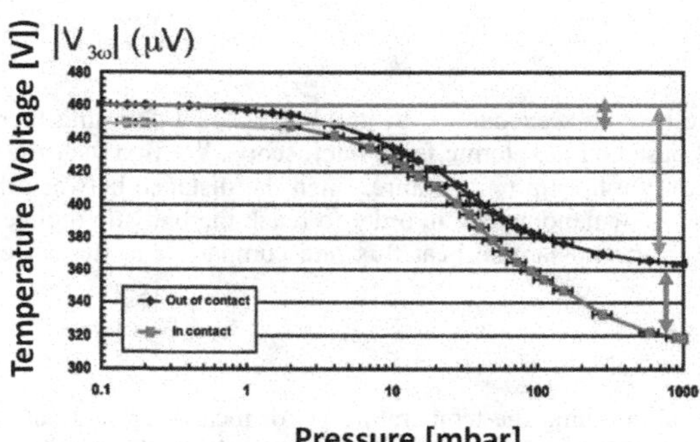


**Figure 1**. Tip voltage $V_{3\omega}$ as a function of the chamber pressure (low vacuum) in the AC mode ($I_0$ = 30 mA and f=70 Hz). (a) Pink line: Tip/sample contact (b) Blue line: Tip removed far away from the sample.

From ambient, the pressure in the vacuum chamber is varied down to approximately 0.1 mbar. Figure 2 shows the third-harmonic tip voltage as a function of the pressure, for two cases: (a) Tip and sample in contact and (b) tip removed far away from the sample (~3 mm). In this so-called 3ω mode [14], the tip's third harmonic voltage $V_{3\omega}$ is proportional to the second harmonic of the temperature $\theta_{2\omega}$: $V_{3\omega} = RI_0/2\ \theta_{2\omega}$, where $R$ is the tip's electrical resistance, $I_0$ the input current's intensity, $\alpha = d\rho_e/dT$ and $\rho_e$ is the Pt90/Rh10 resistivity. At sufficiently low frequency, the temperature field out of the tip is supposed to be in the stationary regime. Arrows underline the temperature differences between the (a) and (b) cases at ambient pressure (orange), between (a) and (b) at low pressure (< 0.1 mbar, green) or between (b) at ambient and (b) at low pressure (grey). At low pressure, the temperature reaches a plateau: it appears that the heat flux does not depend anymore on the pressure. First, we observe that removing the air losses increases the temperature by about 30 % when the probe is in contact and 20 % when out of contact. Second, the heat losses in the air are 2.5 times larger than the one to the sample (grey vs. orange arrows). Finally, the heat flux flowing into the sample is three times more important at ambient pressure than at low pressure (green vs. orange arrows). As a consequence, the heat transfer through the air is the most important channel to lose heat for this kind of tip. This is due in particular to the large size of the tip [14].

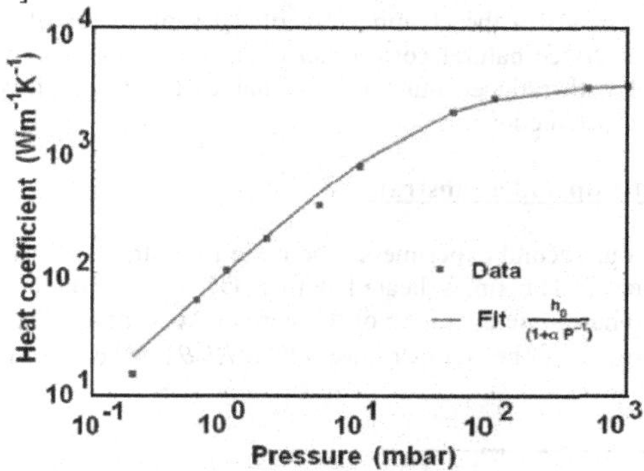


**Figure 2**. Mean heat transfer coefficient (AC mode) vs. pressure. The behavior is linear at low pressures, due to ballistic heat transfer.

The comparison of the experimental values with a model of the temperature in the tip [14, 19] enables to estimate the heat losses from the tip in the air (blue curve). These losses are taken into account by an effective heat transfer coefficient *h* at the fin's walls. This coefficient accounts for heat conduction and convection in air. We find *h(P)* by comparing our model with the measurements (see Fig. 2). The value at the ambient pressure is relatively high *h(*1bar*)*~ 5000 $Wm^{-2}K^{-1}$ as it should be when characteristic sizes are smaller than the convective boundary layer. At low pressures *h(P)* shows a linear dependence. This is due to the ballistic heat transfer between the tip and the apparent walls of our geometry [20] (walls of the microscope, of the sample, of the vacuum chamber). Indeed the mean free path $\Lambda$ of air molecules, which is inversely proportional to the pressure [20], takes values on the order of millimeters at $P = 10^{-1}$ mbar. We compare our experimental data with a model developed by Lees and Liu [20]. They have calculated the heat transfer between two concentric cylinders (radii $R_2 > R_1$) by taking into account the transition between the diffusive and ballistic regimes. They found the following relation for the heat flux:

$$\phi = \frac{\phi_F}{1 + 1 \Big/ \left( 1 + \frac{4}{15} Kn_1^{-1} + \ln \frac{R_2}{R_1} \right)}, \qquad (1)$$

where $\phi_F$ is the corresponding fully diffusive flux and $Kn_1 = \Lambda_1/R_1$ is the Knudsen number associated with cylinder 1 (taken at $T_1$). $\Lambda_1 = (\mu_1/2\rho_1)\ 2\pi m\ k_B T_1$ is the related mean free path, with $\mu$ being the dynamic viscosity, $\rho$ the density, *m* the molecule mass and $k_B$ the Boltzmann constant. Only the density depends on the pressure and we assume that it is inversely proportional. A fit of the data in Figure 2 shows a good agreement between the model by Lees and Liu and the experiment. The transition from diffusive to non-Fourier regimes occurs for a pressure close to $P$~10 mbar. From the fit parameter, assuming a dynamic viscosity equal to $\mu$ = 18 Pa.s, we find an equivalent outer radius of $R_2$ = 10 microns, smaller than the distance to the closer object (microscope head) in the chamber but of the same order of magnitude than the convective boundary layer. Since natural convection is not taken into account in the model by Lees and Liu, this model with reduced outer radius enables to bridge pure conduction in the boundary layer and natural convection.

## Heat transfer between the tip and a substrate

We now describe our second experiment where the tip voltage is measured as a function of the tip-sample distance $d$. The tip is heated with a DC current in order to prevent from intermittent contact due to harmonic dilatation of the wire in AC heating. The tip voltage is again proportional to the average probe temperature ($V=\alpha RI_0\theta$) where $\theta$ is the tip average temperature variation.

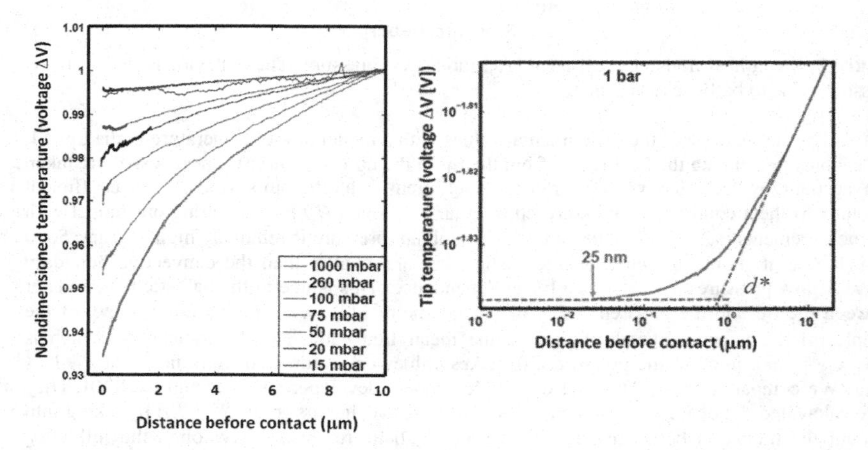


**Figure 3.** Temperature estimated from the tip voltage (DC mode) vs. distance. a) Normalized temperature as a function of various pressures for the last 10 microns before contact. b) Results at ambient pressure (log. scale).

Figure 3a shows the tip voltage divided by its value for $d$=10 μm in the last 10 μm before contact. The distance is measured with the AFM piezo-actuator signal and the contact point is detected thanks to the large deflection of the AFM photodiode when reaching the contact. Note that the absolute distance is plotted here, since the tip attraction by the sample due to various interaction forces has been compensated by monitoring the evolution of the deflection signal out of contact. Before the contact, we observe a steeper slope in the voltage-distance curve meaning that there is a significant tip-sample thermal exchange. Around 6.5% of the temperature is 'lost' at ambient pressure in the last 10 microns. In contrast, when the pressure is lowered, the temperature is maintained within a percent for pressures of a few dozens of millibars.

We observe in Figure 3b the well-known jump that occurs just before contact due to the interaction forces. As a consequence, a domain of distances of 25 nm is not available here. However, a large range of distances can be plotted and the distance scale is logarithmic. The temperature drops linearly for distances larger than 5 μm as shown by the superimposed dashed blue line. The tip temperature appears to level off for distances on the order of 25 nm. The plateau (horizontal dashed line) intercepts the extension of the Fourier regime (dashed blue line) for a characteristic distance $d^*$ ~800 nm.

## DISCUSSION

In order to understand this behavior we compare our experimental data with calculations done for a vertical half torus of radius of curvature $R_c$=15 μm kept at constant temperature, a geometry which is similar to the hot extremity of the tip (not shown here. See [27]). We first verify that heat flux lines can be supposed vertical and the temperature constant in a section of the torus for sufficiently-small values of $d$. By considering that a tip elementary surface element exchanges a flux $\delta\phi=\delta G\Delta T$ with its projection on the sample $\delta S$, with

$$\delta G = \frac{\lambda_0}{d + C\Lambda}\delta S \,, \qquad (2)$$

we observe that a curve with a similar shape to Figure 3b can be reproduced. However, for $C$~1, the characteristic distance $d^*$ is then much smaller than the experimental one, on the order of few tens of nanometers ($d^*_{simulated}$~ 20 nm for a mean free path Λ on the order of 60 nm). Note that in Eq. 2 $\lambda_0$ is the air thermal conductivity at room temperature. Such an expression allows to retrieve the diffusive flux when $d$ is large (i.e. $\delta G= \lambda_0/d$) and a constant conductance when two surfaces are very close, which is typical for the ballistic thermal conduction. It appears that this approach is not sufficient. The roughness of the tip should also be taken into account, as it is of the same order of magnitude than the air molecule mean free path.

## CONCLUSIONS

In conclusion, we have characterized a hot micrometric tip which is used for local thermal analysis. We have shown that a very significant part of the heat transfer between the tip and a sample is due to air heat conduction. Vacuum may be considered as an ideal environment for the use of the SThM, but one should then account for heat exchanges which are very different from the ones at ambient pressure.

## ACKNOWLEDGMENTS

We acknowledge the supports of ANR through the *Monaco* project and of *Fédération Francilienne de Transferts de Masse et de Chaleur*. We thank the members of CNRS research networks "Micro et Nanothermique" and "*Thermique des nanosystèmes et nanomatériaux*" for discussions. We warmly thank J.J Greffet for his help and S.K. Saha for his stay.

## REFERENCES

1. C.C. Williams and H.K. Wickramasinghe, *Appl. Phys. Lett.* 49, (1986), 1587
2. G. B. M. Fiege, A. Altes, R. Heiderhoff and L. J. Balk, *J. Phys. D: Appl. Phys.* 32 , L13 (1999).
3. R.B. Dinwiddie, R.J. Pylkki and P.E. West, Thermal conductivity contrast imaging with a scanning thermal microscope, *Thermal conductivity* 22, T.W. Wong ed, Tecnomics, Lancaster PA, 668-677, 1994
4. A. Hammiche, H.M. Pollock, M. Song, D.J. Hourston, *Rev. Sci. Instrum.* 67, 4268 (1996)
5. P. Vettiger, G. Cross, M. Despont, U. Drechsler, U. Dürig, B. Gotsmann, W. Häberle, M.A. Lantz, H.E. Rothuizen, R. Stutz and G. K. Binnig, *IEEE Trans. on Nanotech.* 1, 39 (2002)
6. W.P. King and K.E. Goodson, ASME *J. Heat Transf.* 124, 597, (2002)
7. W. A. Challener et al. *Nature Photonics* 3,220 (2009).
8. A. Chimmalgi, D.J. Hwang, and C.P. Grigoropoulos, *Nano Lett.* 5, 1924 (2005)
9. A.A. Milner, K. Zhang, and Y. Prior, *Nano Lett.* (2008)
10. O. Fenwick, L. Bozec, D. Credgington, A. Hammiche, G. M. Lazzerini, Y. R. Silberberg and F. Cacialli, *Nature Nanotechnology* 4, 668 (2009)
11. R. Szoszkiewicz, T. Okada, S. C. Jones, T.-D. Li, W. P. King, S. R. Marder, and E. Riedo, *Nano Lett.* 7, 1064 (2007)
12. P.O. Chapuis, J-J. Greffet, K. Joulain and S. Volz. *Nanotechnology* 17, 2978 (2006).
13. Microscale and Nanoscale Heat Transfer, *Topics in Applied Physics* 16, Vol. 107, S. Volz d., 2007
14. S. Lefèvre, S. Volz and P.-O. Chapuis. *Int. J. Heat Mass Transf.* 49, 251 (2006).
15. L. Shi and A. Majumdar, *ASME J. Heat Transf.* 124, 329 (2002)
16. M. Hinz, O. Marti, B. Botsmann, M.A. Lantz and U. Dürig. *Appl. Phy. Lett.* 92, 043122 (2008).
17. S. Lefèvre, S. Volz, C. Fuentes, J.B. Saulnier and N. Trannoy, *Rev. Scient. Instr.* 74, 2418 (2003)
18. S. Lefèvre and S. Volz, *Rev. Scient. Instr.* 76, 033701 (2005)
19. P.-O. Chapuis, *PhD Thesis*, Ecole Centrale Paris (2007)
20. L. Lees and C.Y. Liu, Phys. of Fluids 5, 1137 (1962)

Mater. Res. Soc. Symp. Proc. Vol. 1543 © 2013 Materials Research Society
DOI: 10.1557/opl.2013.675

# Thermal and rheological properties of micro- and nanofluids of copper in diethylene glycol – as heat exchange liquid

Nader Nikkam[1], Morteza Ghanbarpour[2], Mohsin Saleemi[1], Muhammet S. Toprak[1*], Mamoun Muhammed[1] and Rahmatollah Khodabandeh[2]
[1] Department of Materials and Nano Physics, KTH- Royal Institute of Technology, SE-16440 Kista, Stockholm, Sweden.
[2] Department of Energy Technology, KTH- Royal Institute of Technology, SE-100 44 Stockholm, Sweden

## ABSTRACT

This study reports on the fabrication of nanofluids/microfluids (NFs/MFs) with experimental and theoretical investigation of thermal conductivity (TC) and viscosity of diethylene glycol (DEG) base NFs/MFs containing copper nanoparticles (Cu NPs) and copper microparticles (Cu MPs). For this purpose, Cu NPs (20-40 nm) and Cu MPs (0.5-1.5 μm) were dispersed in DEG with particle loading between 1 wt% and 3 wt%. Ultrasonic agitation was used for dispersion and preparation of stable NFs/MFs, and thus the use of surfactants was avoided. The objectives were investigation of impact of size of Cu particle and concentration on TC and viscosity of NFs/MFs on DEG as the model base liquid. The physicochemical properties of all particles and fluids were characterized by using various techniques including Transmission Electron Microscopy (TEM), Scanning Electron Microscopy (SEM) and Dynamic Light Scattering (DLS) techniques. Fourier Transform Infrared Spectroscopy (FTIR) analysis was performed to study particles' surfaces. NFs and MFs exhibited a higher TC than the base liquid, while NFs outperformed MFs showing a potential for their use in heat exchange applications. The TC and viscosity of NFs and MFs were presented, along with a comparison with values from predictive models. While Maxwell model was good at predicting the TC of MFs, it underestimated the TC of NFs, revealing that the model is not directly applicable to the NF systems.

## INTRODUCTION

There is a strong need in cooling industry to develop heat exchange fluids with enhanced thermal properties. Compared to pure conventional heat transfer fluids, such as water or ethylene glycol, suspensions containing solid particles can be a reasonable choice to use as heat exchange fluids because solid particles have higher TC than those of conventional coolants. Stable suspensions with suspended micrometer sized particles (MFs) have already been utilized to enhance the thermal properties of base liquids [1-2]. Recently, nanofluids (NFs) [3], which are novel suspension of nanometric particles, have attracted attention because of their potential to enhance thermal properties of traditional cooling fluids such as water or glycol families. A wide range of materials and base liquids can be combined to fabricate a NF/MF for heat exchange applications. Several groups have reported on NFs / MFs containing a small loading of Cu [4], mesoporous $SiO_2$ [5], CuO [6], Ag [7], CNTs [8] and SiC [9]. There are two major methods for making NFs/MFs: One-step method where particles are formed directly in the base liquid [10], while in the two-step method solid particles are synthesized/acquired and then dispersed in the base liquids [11]. Many factors such as particle size, particle shape, particle loading and base liquid can affect the TC and viscosity of NFs/MFs [12-14]. There are very few studies related to the impact of the particle size in TC and viscosity of NFs/MFs. Our goal is the investigation of impact of Cu NPs and Cu MPs on TC and viscosity of NFs/MFs. There is no report in the literature on DEG containing suspended Cu NPs/MPs as coolant. The reason for selecting DEG as the base liquid is that our study, and discrepancy in the literature, indicated that formulated NFs using low viscosity base liquids exhibited different thermal characteristics making the influence of NPs/MPs inconclusive. In order to understand the real effect of NPs/MPs on the TC, DEG base liquid with higher viscosity than conventional coolants was selected. NFs /MFs were prepared by dispersing commercial Cu NPs/ MPs in DEG with different range of sizes (20-40 nm and 0.5-1.5 μm) at different concentrations and their physicochemical characteristics, including TC and viscosity were evaluated in detail. Moreover, Maxell predictive model was used to estimate the TC of NFs and MFs. The same attempt was also carried out to compare the experimental viscosity data with the predicted values by using Einstein equation.

## EXPERIMENTAL DETAILS

Cu NPs/MPs were purchased from Alfa Aesar, Germany. DEG was obtained from Sigma Aldrich, Germany. In order to study the real effect of NPs/MPs the use of surfactants was avoided. A typical process for making NFs/MFs involves (a) dispersing a known weight of Cu NPs/MPs in DEG as the base liquid; (b) Ultrasonic mixing of the suspension (Chemical instruments AB, CiAB, Sweden) for 25 min. A series of 1wt%, 2wt% and 3wt% Cu NFs/MFs have been prepared. All NFs/MFs were stable for at least 36 h without any visual precipitation.

Scanning Electron Microscopy (SEM) analysis of Cu NPs/MPs was done by using FEG-HR SEM (Zeiss-Ultra 55). Transmission Electron Microscopy (TEM) analysis of the Cu particles size and morphology were performed using JEOL 2100 at 200 kV acceleration. FTIR analysis of solid/liquid samples was performed using Nicolet Avatar IR 360 spectrophotometer, in the range of 400–4000 $cm^{-1}$. Average solvodynamic particle size distribution of Cu NP/MPs was evaluated by Beckmann-Coulter Delsa Nano C system. Finally, TC of NFs/MFs was estimated by using TPS 2500 instrument, which works based on Transient Plane Source (TPS) method. Finally, viscosity of NFs/MFs was evaluated using DV-II+ Pro- Brookfield viscometer.

## RESULTS AND DISCUSSION

### Morphology Analysis

Morphology of Cu NPs and MPs were analyzed by SEM and TEM: micrographs are shown in Figure 1 (a-d). As it can be seen from SEM images in Figure 1(a) and 1(b), Cu NPs and MPs have spherical morphology, with estimated sizes in the range of 20-40 nm and 0.5-1.5 μm, respectively. TEM micrographs of Cu particles, presented in figure 1(c) and 1(d), showed spherical morphology of Cu for both NPs and MPs with the average size of about 20 nm and 0.65 μm for NPs and MPs, respectively. Selected are electron diffraction (ED) patterns of NPs/MPs samples were indexed for Cu phase. Traces of CuO was also observed in ED (figure 1(e) and 1(f)) as well as XRD analyses (data not shown).

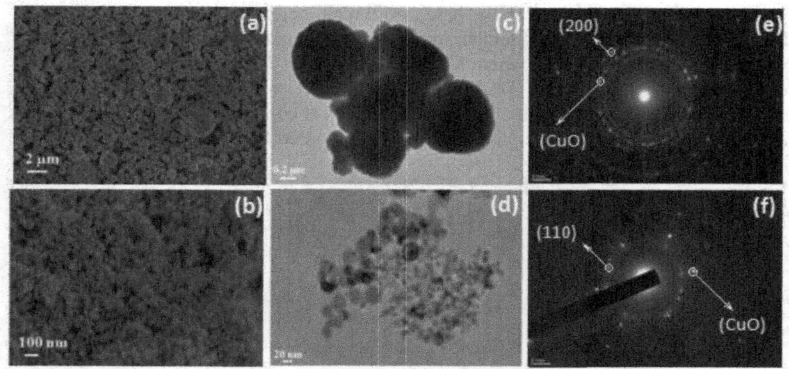


**Figure 1.** SEM micrographs (a, b), (c, d) TEM micrographs and (e, f) electron diffraction patterns of Cu MPs and Cu NPs, respectively.

## Dynamic Light Scattering (DLS) Analysis

It is important to understand the dynamic/dispersed size of NPs/MPs in DEG NFs/MFs. DLS analysis results are presented in Figure 2 for both MFs and NFs. Based on the DLS

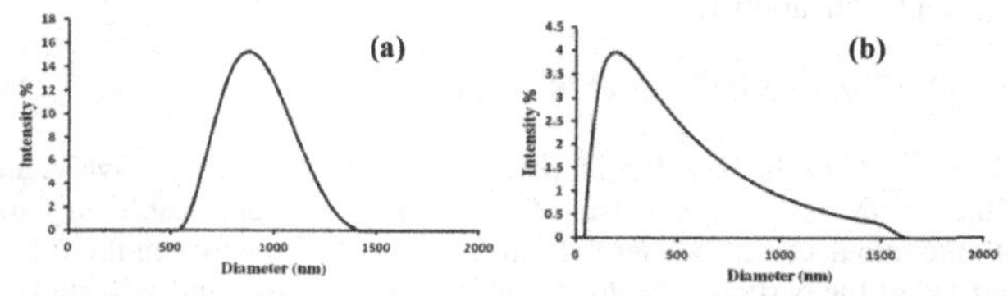


**Figure 2.** Particle size distribution via DLS of: (a) Cu MPs, and (b) Cu NPs in DEG.

analysis, the size of Cu MPs and Cu NPs were estimated in the range of 500-1400 nm and 40-1600 nm with an average solvodynamic size of 900 and 420 nm, respectively. Having such a big difference between primary size obtained from SEM and solvodynamic size measured by DLS shows that Cu NPs/ Cu MPs are agglomerated or aggregated in DEG. In order to distinguish that we performed DLS analysis in water and obtained larger hydrodynamic size than that in DEG. This reveals the aggregation of both Cu NP and Cu MP probably due to their preparation process, while DEG dispersion helped to break agglomerates.
In order to study the surface of NPs/MPs, FTIR analysis was performed. The results revealed that NPs/MPs surfaces have been modified by DEG base liquid, which helps to stabilize the particles in DEG media (data not shown).

## TC measurements of NFs/MFs

Several DEG base NFs/ MFs with various Cu NPs/Cu MPs content from 1 wt% to 3 wt% were evaluated for their TC using TPS method, and results are presented in Figure 3(a).

NFs/MFs exhibit higher TC values than the base liquid with increasing loading of Cu NPs/Cu MPs in DEG and also with increasing temperature. Figure 3(b) exhibits effective TC ($K_{nf}$ and $K_{bl}$) of NFs/MFs at 20 °C and 40 °C, where $K_{nf}$ and $K_{bl}$ stand for TC of NFs/MFs and the base liquid, respectively. The maximum TC enhancement of ~ 2.8 % was obtained for Cu NFs with 3wt% Cu NPs at 20 °C. Moreover, Cu NFs showed higher TC compared to the Cu MFs, which may be attributed to the higher surface area of Cu NPs compared to Cu MPs.

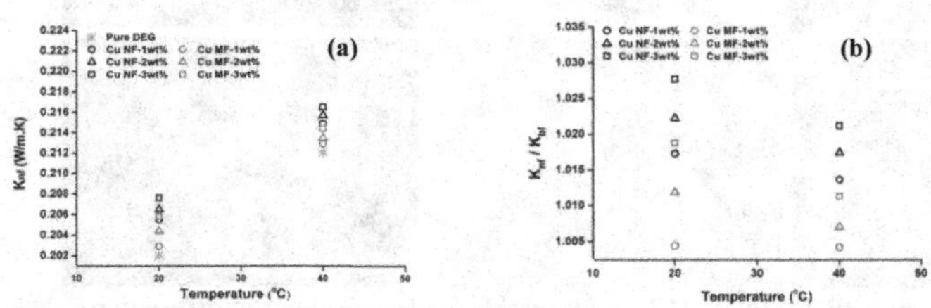


**Figure 3.** TC values of NFs/MFs at different temperatures (a) with various Cu NPs/Cu MPs concentration and (b) Effective TC of NFs/MFs at different Cu concentration and temperatures.

## Comparison between experimental results and predictive models

Maxwell effective medium theory [15] has been used in order to estimate TC of formulated NFs / MFs and compare with the measurement results. The TC ratio predicted by the Maxwell effective medium theory is given by equation 1:

$$K_{nf/mf}/K_{bl}= K_P+2K_{bl} -2\varnothing\ (K_{bl}-K_p)\ /\ KP+2K_{bl}+\varnothing\ (K_{bl} -K_p) \quad (1)$$

wherein, $K_{bl}$, $K_p$, and $K_{nf/mf}$ are the TC of the base liquid, particle, and NF/MFs, respectively, and $\varnothing$ is the volume fraction of the NPs/MPs. This expression for effective TC is applicable only for spherical particles, does not take into account the thermal interface resistance between the liquid and particles and the reduced TC of the particles due to size effects. Figure 4(a) and 4(b) show the TC of NFs/MFs for both experimental and predicted values by Maxwell model in the weight fractions ranging from 1% to 3% at 20 °C and 40 °C.

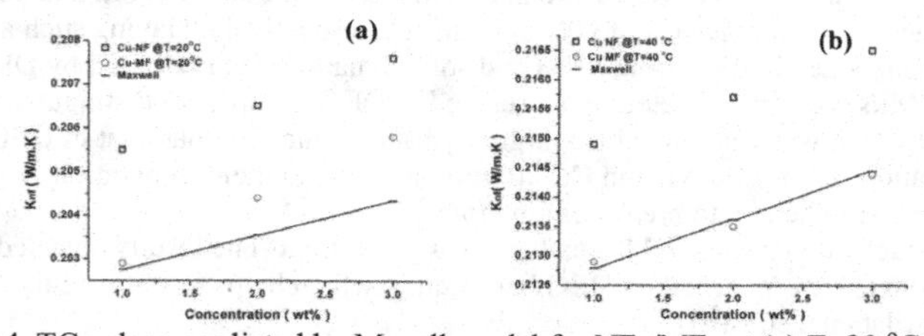


**Figure 4.** TC values predicted by Maxell model for NFs/MFs at (a) T=20 °C and (b) T=40 °C

Maxwell model could estimate the TC of MFs quite well at 40 °C while TC of NFs with Cu NPs is significantly higher than the Maxwell values. This indicates the important role of the interface between the liquid and particles that have been neglected in the Maxwell model.

## Rheological Evaluation

The viscosity of the NFs/MFs has essential role in thermophysical properties of the suspensions [16]. Then viscosity measurements were carried out in order to investigate the rheological properties of NFs/ MFs with particle loading ranging from 1wt% to 3wt% at 20 °C. All NFs/MFs exhibited Newtonian behavior (Figure not shown), which means the viscosity as a function of shear rate is constant. The results presented in Figure 5(a) show that viscosity of DEG, and of the NFs/MFs, increased as the loading of Cu particle was increased. Figure 5(b) presents the ratio of viscosity of NFs/MFs ($\mu_{nf}$) to the viscosity base liquid ($\mu_{bl}$) at 20 °C. The figure shows that NFs containing Cu NPs has higher increase in viscosity compared to the MFs with Cu MPs, which can be attributed to the large contact area between Cu NPs and DEG base liquid.

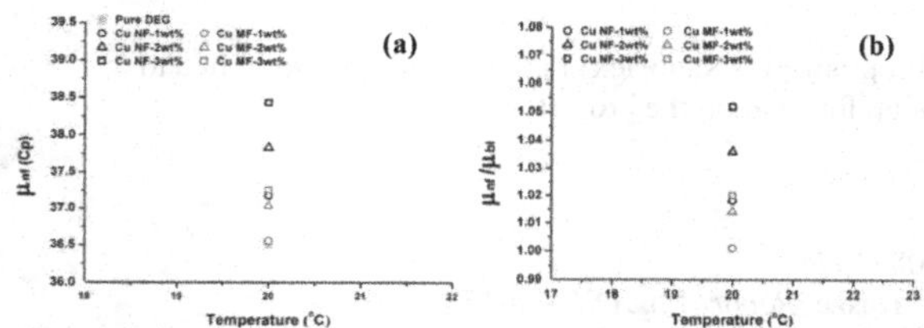


**Figure 5**. (a) Viscosity values of DEG base NFs/MFs with various Cu NPs/MPs concentrations at 20 °C.(b) The viscosity ratio of NFs/MFs to the viscosity of DEG as base liquid at 20 °C

## Comparison between experimental viscosity results and predictive model

Einstein law of viscosity [17] was used for predicting effective viscosity of NFs / MFs according to equation (2).

$$\mu_{nf}/\mu_{bl} = 1 + 2.5\,\varnothing \qquad (2)$$

Where $\mu_{bl}$ and $\mu_{nf}$ are the viscosity of the base liquid and the NF/MF, respectively. Both the measured and predicted viscosity ratios are plotted in Figure 6. The Einstein equation can estimate quite well the viscosity of MFs containing Cu MPs while it underestimates the viscosity of NFs, showing the limitation of the equation's applicability to the NF systems for viscosity prediction.

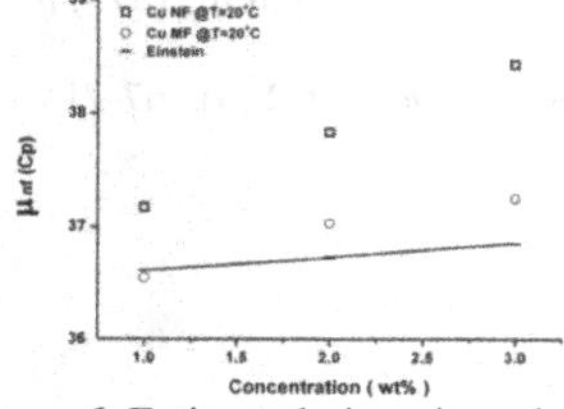


**Figure 6.** Estimated viscosity values from Einstein equation for NFs/MFs at 20 °C.

## CONCLUSIONS

We presented on the fabrication and evaluation of DEG base NFs/MFs containing Cu NPs/MPs for heat exchange and reported an enhancement of TC of the base liquid due to the presence of Cu NPs/MPs. SEM analysis revealed spherical particle morphology with an average size of 20-40 nm and 05-1.5 μm for NPs and MPs, respectively. The stable NFs dispersion has been prepared by a two-step method, using high power ultrasonic agitation. The NFs presented in this study exhibited higher TC values than the base liquid and MFs, showing a potential for their use in heat exchange applications. While Maxwell model was good at predicting the TC of MFs, it underestimated the TC of NFs, revealing that the model is not directly applicable to the NF systems.

## ACKNOWLEDGMENTS

This study is supported by FP7 Program EU (NanoHex) Project. Authors would like to acknowledge the European Commission for funding the project.

## REFERENCES

1. A. S. Ahuja, *J. Appl. Phys.* **46**, 3408 (1975)
2. K. V. Liu, S. U. S. Choi, and K. E. Kasza, *Report No. ANL* -88-153, 1988.
3. S.U.S. Choi, *ASME FED*, **231**, 99(1995).
4. Y. Xuan and Q. Li, *Int. J. Heat Fluid Flow,* **21**, 58 (2000).
5. N. Nikkam, M.Saleemi, S.Li, M.Toprak, E.B.Haghighi, R. Khodabandeh and B.Palm, *J. Nanopart. Res.* **13** (11), 6201-6 (2011).
6. S. Lee, S.U.S. Choi, S. Li and J.A. Eastman, *ASME J. Heat Transfer*, **121**, 280 (1999).
7. P. Sharma, I. Baek, T. Cho, S. Park and K.B. Lee , *Powder Technology*, **208**, 7 (2011).
8. S. U. S. Choi, Z.G. Zhang, W. Yu and F.E. Lockwood, E.A. Grulke, *Appl. Phys. Lett.* **79**, 2252 (2001).
9. H. Q. Xie, J.C. Wang, T.G. Xi andY. Liu, *J. Chin. Ceram.* Soc. **29**, 361(2001).
10. J. A. Eastman, S. U. S. Choi, S.Li, W. Yu and L. J.Thompson, *Appl. Phys. Lett.* **78**, 718 (2001).
11. S. U. S. Choi, *J. Heat Transfer,* **131**, 033106 (2009).
12. S. H. Kim, S. R. Choi and D. Kim, *ASME Trans. J. Heat Transfer* **129**, 298 (2007)
13. W. Yu, D.M. France and S.U.S. Choi, *J.L. Heat Transfer Eng.* **29** (5), 432–460 (2008).
14. M. Chopkar, S. Sudarshan, P.K. Das and I. Manna, *Metallurgical and materials transactions A*, **39A**, 1535(2008).
15. J. C. Maxwell, *A Treatise on Electricity and Magnetism*, (Clarendon Press, Oxford, 1873) p. 365
16. P. K. Namburu, D.P. Kulkarni, A. Dandekar and D.K. Das, *Micro Nano Lett.* **2** (3), 67-71 (2007).
17. T. S. Chow, *Phys. Rev.* **E 48**, 1977 (1993).

Mater. Res. Soc. Symp. Proc. Vol. 1543 © 2013 Materials Research Society
DOI: 10.1557/opl.2013.939

# Carrier Mapping in Thermoelectric Materials

Georgios S. Polymeris[1], Euripides Hatzikraniotis[1], Eleni C. Stefanaki[1], Eleni Pavlidou[1], Theodora Kyratsi[2], Konstantinos M. Paraskevopoulos[1], Mercouri G. Kanatzidis[3]
[1]Physics Department, Aristotle University of Thessaloniki, GR- 54124, Thessaloniki, Greece
[2] Department of Mechanical and Manufacturing Engineering, University of Cyprus, 1678 Nicosia, Cyprus.
[2] Department of Chemistry, Northwestern University, 2145 North Sheridan Road, Evanston, IL 60208, U.S.A

## ABSTRACT

The application of micro-fourier transform infrared (FTIR) mapping analysis to thermoelectric materials towards identification of doping inhomogeneities is described. Micro-FTIR, in conjunction with fitting, is used as analytical tool for probing carrier content gradients. The plasmon frequency $\omega_P{}^2$ was studied as potential effective probe for carrier inhomogeneity and consequently doping differentiation based on its dependence of the carrier concentration. The method was applied to PbTe-, PbSe- and $Mg_2Si$- based thermoelectric materials.

## INTRODUCTION

In the past decade several promising bulk thermoelectric materials have been identified and developed, including filled skutterudites [1], complex bismuth tellurides [2], nanostructured lead chalcogenides such as PbTe and PbSe [3-8], magnesium silicides [9] and Zn-Sb alloys [10]. In any of the aforementioned cases, a microscopically homogeneous material has to be assumed in order to acquire a comprehensive understanding between the measured thermoelectric properties, such as the Seebeck coefficient and the microscopic intrinsic features of each material.

One of the promising ways of increasing the thermoelectric efficiency is the development of new inhomogeneous materials and structures including nanostructuring, heterostructures, segmented and functionally graded materials [7]. In functionally graded materials with carrier concentration gradient, the thermoelectric properties vary continuously along the length due to a continuous compositional or doping gradient, whereas in segmented materials the thermoelectric properties are changed step-like at the interface where two dissimilar materials are bonded together. Another approach of introducing in-homogeneities is the doping modulation. Usually, thermoelectric materials are heavily doped semiconductors, and a guest element is used to tune the carrier concentration with a reduction of carrier mobility due to a notable ionized impurity-electron scattering [11]. Modulation doped materials are two-phase composites with a matrix-phase of low carrier concentration and heavily doped inclusions used to provide the carriers [12].

Free carrier concentration is thus, a key parameter in thermoelectric materials. Intentional or unintentional phase and doping separations create local variations in the material composition and therefore in thermoelectric parameters. Accurate profiling is consequently becoming increasingly important and a number of profiling techniques are available today for spatial resolution of thermal and electrical properties, namely, resistance thermometry [13], scanning thermal microscopy [14], near-field [15] and infrared thermometry [16]. Mapping of in-homogeneities has been monitored by microscopic Seebeck coefficient distribution [17] in materials such as $Bi_2Te_{2.85}Se_{0.15}$ [18,

19], both zone melted and hot pressed samples of an iodine doped $Bi_2Te_{2.85}Se_{0.15}$ based alloy [20], as well as n-type $AgPb_{18}SbTe_{20}$ and p-type $AgPb_9Sn_9SbTe_{20}$ [21].

The current work describes the application of a non-destructive, reflection-based, mid infrared micro-spectroscopic mapping analysis into thermoelectric materials towards identification of position-sensitive structural in-homogeneities. Dopant in-homogeneities were also directly confirmed by SEM-EDX. Our approach is based on the dependence of the plasmon frequency $\omega_P^2$ on the carrier concentration.

## EXPERIMENTAL

Room temperature infrared spectroscopic measurements in the reflectivity configuration were performed with near normal incidence light in the range of 500 – 4000 $cm^{-1}$ using the microscope Perkin Elmer i-series, connected with an FTIR spectrometer. Conventional IR reflectivity measurements are usually carried out with an iris of about 2 mm to 4 mm in diameter. In micro-IR, the iris is reduced to 100 μm, which enables FTIR mapping of the sample for local in-homogeneities throughout a pre-selected area of the sample's surface. The sample was placed on the holder and an image survey was taken for the depiction of the area under investigation. The spatially resolved spectral information was obtained by scanning the sample spot by spot across the surface using a stationary detector and collecting at each spot a complete infrared spectrum. The sample could be moved by an x-y computer controlled stage to position the next spot "under" the microscope, at a spatial resolution of 500 μm. The reflection coefficient was determined by typical sample in/sample out method with a gold mirror as reference.

Each obtained experimental reflectivity spectrum $R(\omega)$, i.e. the response of the material to the incident electromagnetic radiation, is characterized by a complex dielectric function (complex permittivity). The complex dielectric function $\tilde{\varepsilon}(\omega)$ is related to the measured reflectivity via the relation:

$$R(\omega) = \left| \frac{\sqrt{\tilde{\varepsilon}(\omega)} - 1}{\sqrt{\tilde{\varepsilon}(\omega)} + 1} \right|^2 \quad \text{with} \quad \tilde{\varepsilon}(\omega) = \varepsilon_\infty - \frac{\varepsilon_\infty \cdot \omega_P^2}{\omega^2 + i \cdot \gamma_P \omega} \quad \text{and} \quad \omega_P^2 = \frac{N \cdot e^2}{\varepsilon_0 m_e^* \varepsilon_\infty}, \quad (1)$$

where $\varepsilon_0$ is the vacuum permeability, $m_e$* is the free carriers conductivity effective mass, $e$ is the electron charge and $N$ the free carrier concentration. Each spectrum was fitted by applying the conventional Drude model (eq. 1), which describes the interaction between the free carriers and the infrared radiation. The main outcome of this fitting procedure involves the values for the three fitting parameters $\omega_P$, being the plasmon frequency, $\gamma_P$, the free carrier damping factor and the high frequency dielectric constant, $\varepsilon_\infty$.

Due to the dependence of the carrier concentration ($N$) on the $(\omega_P)^2$, the present study exploits the usefulness of the experimentally evaluated $\omega_P$ parameter, obtained after fitting, as a measure of the doping inhomogeneity of the thermoelectric samples. These inhomogeneities are taken to be large in the sense that each point in the material can be associated with a different macroscopic free carrier concentration and consequently a different macroscopic $\omega_P$ parameter.

For the direct doping gradient identification, the chemical composition and surface condition of the samples were determined by Energy dispersive X-ray (EDX) and SEM analyses using a Jeol 840A scanning microscope with an energy-dispersive spectrometer attached (model ISIS 300; Oxford) for energy-dispersive x-ray analysis (EDX).

The samples of the present study were the following: (a) p-type PbSe crystal doped with 0.4% Na, (b) p-type PbTe crystal doped with 1.6% Br, (c) $Mg_2Si$ crystal doped with 4% Sb. All crystals were of circular shape (pellet); therefore the mapping took place throughout a rectangular area of dimensions 4x4 mm and 3.5x3.5 mm of each surface respectively at a spatial resolution of 500 μm.

## RESULTS

The mapping procedure allows for the acquisition of infrared spectra with spatial information. Since the major result of the fitting procedure deals mostly with the fitting parameter $\omega_P$, the primary information derived from the mapping procedure is presented as a contour plot of the plasmon frequencies for each sample; these are presented in Figure 1 for all three samples. The configuration of these contours allows inferring the relative gradient of the plasmon frequency parameter and estimating its value at specific places all around the surface, providing spectroscopic and spatially resolved information simultaneously.

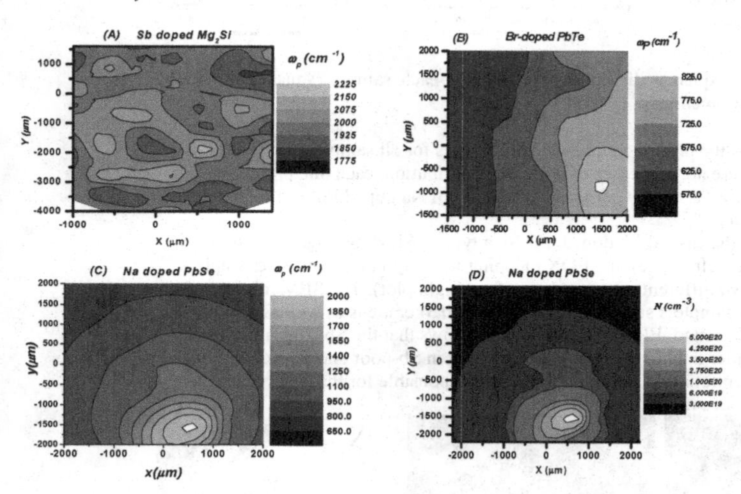


**Figure1**. Contour $\omega_p$ plots for the Sb doped $Mg_2Si$ (*A*) , Br doped PbTe (*B*) and Na doped PbSe (*C*). Plot (*D*) presents the contour *N* plot for the latter sample.

Note that the coloring scale of these contour plots is different because: (a) the different plasma frequencies monitored for each one of the samples and (b) the much larger scatter in the plasma frequency values for the case of PbSe. These contours provide direct evidence of the inhomogeneity of the samples subjected to the present study. Inhomogeneity was probed in all three samples. One parameter that differs strongly among these samples is type and the gradient magnitude. For Sb-doped $Mg_2Si$ sample, scattered dopant-rich areas are observed, and the sample resembles more a modulation doped material. The other two samples (Na doped PbSe and Br doped PbTe) appear more like functionally graded with dopant concentration gradients. The gradient, being always perpendicular to the contour lines, yields large magnitude when the lines are close together: and the variation is steep. For lead chalcogenide samples, it seems that there is an area yielding maximum $\omega_P$ values, however as the distance of the studied points from this area increases, the $\omega_P$ values are decreasing in radial form. This feature is more prominent in the case of the Na doped PbSe. The bottom right plot of Figure 1 presents the contour plots of the carrier concentration for Na doped PbSe. The effective mass $m_e$*

was found to increase with increasing hole density [22]. As it becomes obvious, the patterns of both contours are identical due to the $(\omega_P)^2$ dependence of $N$.

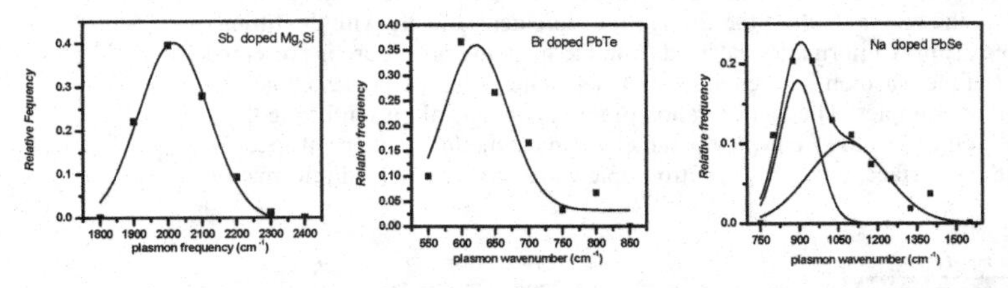


**Figure 2.** Plasmon frequency distribution profile for each sample (squares), fitted by Gaussian distribution functions.

Figure 2 presents the plasma frequency distributions for all samples. Except for one notable exception, there are two cases of unimodal distribution, each one presenting one peak which can be fit by a single Gaussian. In the case of Na doped PbSe, a bimodal distribution was observed.

In Figure 3, for the case of Sb doped $Mg_2Si$, a typical SEM micrograph is shown in the backscatter mode (left image), the EDX mapping for Sb (middle) and an example of spectra taken from two different spots on the surface (right plot). The SEM micrograph depicts details of the sample's surface; as can be seen, darker areas as well as white spots appear in a gray background. EDX mapping clearly shows that the Sb concentration in the dark areas is smaller than in the gray area, resulting in Sb-poor and Sb-rich regions [23]. The local concentration of Sb-rich regions are responsible for the observed areas with higher carrier concentration.

**Figure 3**. SEM micrograph, EDS dopant mapping (white spots) and IR spectra taken from two different spots on the surface (gray and darker region), for Sb doped $Mg_2Si$

For the case of Na doped PbSe, the area corresponding to maximum $\omega_P$ values coincides with the presence of a large mechanical hole on the surface of the sample. The gradual variation of the plasmon frequency takes place more prominently compared to the cases of the other two samples. This specific type of in-homogeneity that was monitored could, possibly, be attributed to a doping gradient on the sodium content. Assuming this is the case; the dependence of the latter on the $\omega_P$ parameter could provide an indirect way for calibrating the doping concentration and thus, use the $\omega_P$ parameter as a measure of the sodium content. In order to further investigate this previous statement, additional EDX measurements were performed in the case of this specific sample along the Y-axis which passes through the area yielding the maximum plasmon frequency (i.e. the mechanical hole), in steps of 500 μm. Figure 4a presents at the same plot the variation of $\omega_P$ parameter values obtained by the FTIR mapping procedure as well as the sodium content obtained by the EDX analysis along the Y axis. As this Figure reveals, those two, independently measured physical quantities indicate the same systematic trend along the Y-axis of the sample. Consequently, these two quantities could be correlated. Moreover, it is safe to assume that this correlation is valid not only along the specific Y-axis of the measurements but throughout the surface of the sample.

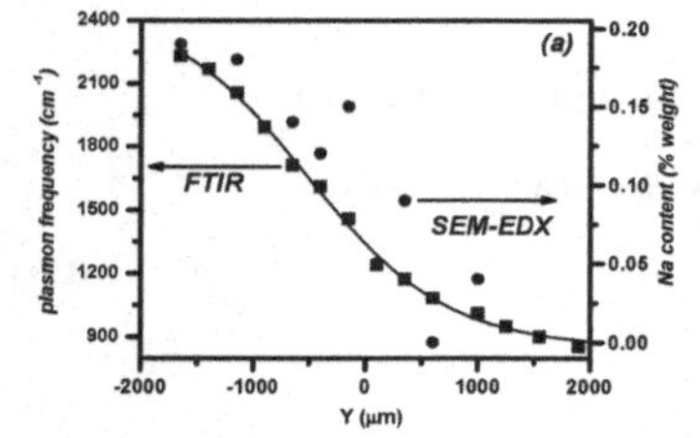


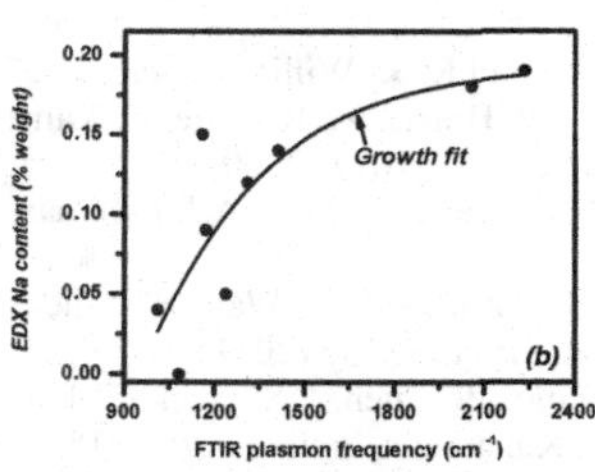


**Figure 4.** Plot (a) presents the variation of the plasma frequencies (squares, left-hand Y-axis) as well as of the Na content (dots, right-hand Y-axis) across a Y-axis from the area with maximum $\omega_P$, for the Na doped PbSe crystal. Right plot depicts the correlation of these two physical quantities along with the corresponding polynomial fit (straight line).

Figure 4b presents the plot of the sodium content obtained by the EDX analysis versus the plasmon frequency obtained for the measurements of the micro-FTIR (dots) along with the corresponding polynomial fitted line. This results in the $(\omega_P)^2$ dependence of the plasma frequency on $N$, and the super-linear dependence of the conductivity effective mass [22]. Based on this correlation, each plasmon frequency value obtained corresponds to a specific value of Na content, thus demonstrating the quantitative power of our micro-FTIR doping spectroscopy. It is notable from Figure 4, that in the case of lower values of sodium, namely lower than 0.05 at % weight the SEM-EDX technique presents difficulty in detecting chemical variations. In contrast, the micro-FTIR mapping analysis seems much more sensitive in discriminating between low levels of dopans. This better sensitivity is attributed to the better resolution of the $\omega_P$ parameter value, being of the order of 20 $cm^{-1}$.

## CONCLUSIONS

The application of micro-FTIR mapping, in conjunction with minimization techniques in fitting experimental data, is a powerful approach for the identification of position-sensitive structural and electronic inhomogeneities in semiconductor samples. The present study provides with the first promising results using the plasmon frequency $\omega_P$ as a probe for detecting spatial compositional variations affected by the doping concentration with spatial resolution of 0.5 mm. The presence of these variations was also verified by SEM-EDX elemental analysis but not when the concentrations involved are low. The micro-FTIR mapping technique offers a credible, simple and convenient way to assess the electronic doping in large bulk samples of the thermoelectric mateirals. Further work is required in order to improve the spatial resolution of the technique.

## ACKNOWLEDGMENTS

This work is partially supported from the ThermoMag Project, which is co-funded by the European Commission in the 7th Framework Programme (contract NMP4-SL-2011-

263207), by the European Space Agency and by the individual partner organizations. MGK acknowledges support from the Center for Revolutionary Materials for Solid State Energy Conversion, an EFRC funded by the U.S. Department of Energy, Office of Science, Office of Basic Energy Sciences under Award DESC0001054

## REFERENCES

1. B. C. Sales, D. Mandrus and R. K. Williams, *Science 272*, 1325 (1996).
2. D-Y. Chung, T. Hogan, P. Brazis, M. R. Lane, C. Kamewurf, M. Bastea, C. Uher and M. G. Kanatzidis, *Science 287*, 1024 (2000).
3. H. Peng, J-H. Song, M.G. Kanatzidis and A.J. Freeman, *Phys. Rev. B 84*, 125207 (2010).
4. K. Biswas, J. Q. He, Q.C. Zhang, G. Y.Wang, C. Uher, V. P. Dravid, M. G. Kanatzidis, *Nature Chemistry 3*, 160 (2011).
5. K. F. Hsu, S. Loo, F. Guo, W. Chen, J. S. Dyck, C. Uher, T. Hogan, E. K. Polychroniadis, M. G. Kanatzidis, *Science 303*, 818 (2004).
6. M. G. Kanatzidis, *Recent Trends in ThermoelectricMaterials Research I 69*, 51 (2001).
7. M. G. Kanatzidis, *Chemistry of Materials 22*, 648 (2010).
8. J. R. Sootsman, D. Y. Chung, M. G. Kanatzidis, *Angewandte Chemie-International Edition, 48*, 8616 (2009).
9. M.I. Fedorov and V.K. Zaitsev, in *Thermoelectrics and its Energy Harvesting*, ed D.M. Rowe, Taylor & Francis, 11-1 (2012)
10. T. Caillat, J.-P. Fleurial and A. Borshchevsky, *J. Phys. Chem. Solids 58*, 1119 (1997).
11. W. Liu, X. Yan, G. Chen, Z. Ren, *Nano Energy 1*, 42 (2012).
12. M. Zebarjadi, G. Joshi, G.H. Zhu, B. Yu, A. Minnich, Y.C. Lan, X.W. Wang, M. Dresselhaus, Z.F. Ren, G. Chen, *NanoLetters 11*, 2225 (2011).
13. K. E. Goodson, M. I. Flik, L. T. Su, D. A. Antoniadis, *J. Heat Transfer 117*, 574 (1996).
14. A. Majumdar, J. P. Carrejo, *J. Lai, Appl. Phys. Lett. 62*, 2501 (1993).
15. B. D. Boudreau, J. Raja, R. J. Hocken, S. R. Patterson, *J. Patten, Rev. Scie. Instr. 68*, 3096 (1997).
16. D.A. Fletcher, D.S. Kino, K.E. Goodson, *Microscale Thermophysical Engin. 7*, 267 (2003).
17. S. Iwanaga, J. Snyder: *J. Electr. Mat. 41,* 1667 (2012)
18. T. E. Svechnikova, P . P . Konstantinov, M. K. Zhitinskaya, S . A. Nemov, D. Platzek and E. Muller, *Proceedings of the 7th European Workshop on Thermoelectrics* Spain, (2002)
19. D. Platzek, A. Zuber, C. Stiewe, G. Bahr, P . Reinshaus and E. Muller, in Proc. Int. Conf. on Thermoelectrics, La Grade-Motte, France, 2003, IEEE, Piscataway, NJ 08855, USA, 528.
20. H.L. Ni, X.B. Zhao, G. Kaprinski, E. Muller: *Journal of Materials Science 40*, 605 (2005)
21. A. Kosuga, K. Kurosaki, H. Muta, C. Stiewe, G. Kaprinski, E. Muller, S. Yamanaka: *Materials Transactions 47*, 1440 (2006).
22. Th.C. Chasapis, Y. Lee, G.S. Polymeris, E-C. Stefanaki, E. Hatzikraniotis, X. Zhou, C. Uher, K.M. Paraskevopoulos, M.G. Kanatzidis: *MRS Proceedings*, 1490 DOI: http://dx.doi.org/10.1557/opl.2013.150 (2013).
23. M. Ioannou, G. Polymeris, E. Hatzikraniotis, A.U. Khan, K.M. Paraskevopoulos, Th. Kyratsi, *Journal of Electronic Materials*, DOI: 10.1007/s11664-012-2442-6 (2013).

# AUTHOR INDEX

Abe, H., 57
Albaret, Tristan, 71
Antohe, S., 125
Assy, Ali, 159

Barr, J.A., 39
Becker, A., 3
Beckman, Scott P., 23, 39
Blandre, Etienne, 71

Cammarata, Robert C., 113
Cao, Helin, 9
Chan, Tsung-ta E., 93
Chapuis, Pierre-Olivier, 159
Chavez, R., 3
Chen, Yong P., 9
Chi, Su (Ike) Chih, 113
Chien, L-C, 13
Chopra, Nitin, 119
Coleman, Elane, 131

Engenhorst, Markus, 3, 99

Farias, Stephen L., 113
Farva, Umme, 137
France-Lanord, Arthur, 71
Fryauf, David M., 131

Gautam, Devendraprakash, 99
Ghanbarpour, Morteza, 143, 165
Gomès, Séverine, 159
Greaney, P. Alex, 65
Green, Andrew J., 151
Gu, Haiming, 13

Hagino, H., 57
Hatzikraniotis, Euripides, 171
Honda, H., 57

Ichinose, A., 57
Ioannidou, Alexandra, 29
Ion, L., 125

Jacquot, Alexandre, 105
Jägle, Martin, 105
Jean, Valentin, 71

Kanatzidis, Mercouri G., 171
Karvonen, Lassi, 83
Kessler, V., 3
Khodabandeh, Rahmatollah, 143, 165
Kobayashi, Nobuhiko P., 131
Koch, Carl C., 93
Kyratsi, Theodora, 171

Lacroix, David, 71
LeBeau, James M., 93
Lefèvre, Stéphane, 159
Levin, George A., 43
Li, Xinyu, 13
Li, Yuan, 119
Lu, S.G., 13

Matsumoto, K., 57
Mele, P., 57
Merabia, Samy, 71
Miotkowski, Ireneusz, 9
Mitran, T.L., 125
Miyazaki, K., 57
Mogare, Kailash, 83
Muhammed, Mamoun, 105, 143, 165

Nemnes, G.A., 125
Niarchos, Dimitrios G., 29
Nicolaev, Adela, 125
Nikkam, Nader, 143, 165
Nishimatsu, T., 39
Norris, Kate J., 131

Oliveira, Laura de Sousa, 65

Panasyuk, George Y., 43
Paraskevopoulos, Konstantinos M., 171
Park, Chan, 137
Pavlidou, Eleni, 171
Petermann, N., 3
Polymeris, Georgios S., 171
Populoh, Sascha, 83

Qian, Xiao-Shi, 13

Richardson, Hugh H., 151
Rousseau, Emmanuel, 159

Sagarna, Leyre, 83
Saini, S., 57
Saleemi, Mohsin, 105, 165
Saucke, Gesine, 83
Schierning, Gabi, 3, 99
Schilling, Carolin, 99
Schmechel, Roland, 3, 99
Singh, Sathya P., 143
Srivastava, Gyaneshwar P., 49
Stefanaki, Eleni C., 171
Stuart, Judy, 93

Tafti, Mohsen Y., 105
Tarkhanyan, Roland H., 29
Termentzidis, Konstantinos, 71
Thomas, Peter, 93
Tompa, Gary S., 131
Toprak, Muhammet S., 105, 143, 165

Venkatasubramanian, Rama, 93
Visan, Camelia, 125
Vogel-Schäuble, Nina, 83
Volz, Sebastian, 159

Wan, Liwen F., 23
Weidenkaff, Anke, 83
Wiggers, H., 3
Winterer, Markus, 99

Xu, Yang, 9

Yelgel, Ö. Ceyda, 49
Yerkes, Kirk L., 43

Zhang, Junce, 131
Zhang, Q.M., 13

# SUBJECT INDEX

amorphous, 71
annealing, 119, 137
Au, 151

c, 65
calorimetry, 159
ceramic, 39
chemical synthesis, 105
cryomilling, 93
crystal, 23
crystal growth, 131
crystalline, 137
Cu, 131

dielectric properties, 13
differential thermal analysis (DTA), 159

electrical properties, 13
electrodeposition, 113
electronic material, 23
Er, 151

ferroelectric, 39
fluid, 143, 165

infrared (IR) spectroscopy, 171

liquid crystal, 13

metalorganic deposition, 131

nanostructure, 3, 29, 43, 93, 99, 105, 143, 165

oxide, 39, 57, 83, 119

Pb, 171
porosity, 29

scanning electron microscopy (SEM), 137
semiconducting, 9, 99
si, 71
simulation, 43, 65
spintronic, 125

thermal conductivity, 43, 49, 65, 71, 83, 119, 143, 151, 159, 165
thermoelectric, 3, 9, 23, 49, 83, 93, 105, 113, 171
thermoelectricity, 3, 9, 29, 49, 57, 99, 125
thin film, 57, 113
Tl, 9

Printed in the United States
by Baker & Taylor Publisher Services